AF847116

Report on Yangtze River Rehabilitation and Protection 2019

Changjiang Water Resources Commission of
Ministry of Water Resources of the People's
Republic of China

Report on Yangtze River Rehabilitation and Protection 2019

长江出版社
CHANGJIANG PRESS

Springer

Changjiang Water Resources Commission of
Ministry of Water Resources of the People's
Republic of China
Wuhan, China

ISBN 978-981-16-4926-4 ISBN 978-981-16-4927-1 (eBook)
https://doi.org/10.1007/978-981-16-4927-1

Jointly published with Changjiang Press (Wuhan) Co., Ltd.
The print edition is not for sale in China (Mainland). Customers from China (Mainland) please order the print book from: Changjiang Press (Wuhan) Co., Ltd.

© Changjiang Press (Wuhan) Co., Ltd. 2022
This work is subject to copyright. All rights are solely and exclusively licensed by the Publisher, whether the whole or part of the material is concerned, specifically the rights of reprinting, reuse of illustrations, recitation, broadcasting, reproduction on microfilms or in any other physical way, and transmission or information storage and retrieval, electronic adaptation, computer software, or by similar or dissimilar methodology now known or hereafter developed.
The use of general descriptive names, registered names, trademarks, service marks, etc. in this publication does not imply, even in the absence of a specific statement, that such names are exempt from the relevant protective laws and regulations and therefore free for general use.
The publishers, the authors, and the editors are safe to assume that the advice and information in this book are believed to be true and accurate at the date of publication. Neither the publishers nor the authors or the editors give a warranty, express or implied, with respect to the material contained herein or for any errors or omissions that may have been made. The publishers remain neutral with regard to jurisdictional claims in published maps and institutional affiliations.

This Springer imprint is published by the registered company Springer Nature Singapore Pte Ltd.
The registered company address is: 152 Beach Road, #21-01/04 Gateway East, Singapore 189721, Singapore

Editorial Board

Chairman

Jianhua Ma

Vice Chairman

Jiajun Hu
Zhiyu Zhong
Weizhong Hu

Members of the Board

Shuishan Ma
Fengcong Rong
Guiya Chen
Xiaoyong Luo
Qihui Yu
Zhaoming Xu
Wenyao Mao
Yingxi Li
Bing Tang
Sunyun Lv
Jionghong Chen
Xiang Li
Wei Yao
Hongbo Song
Fang Shi

Shuangxi Li
Peng Wang
Shengxiang Ding
Guangcai Chen
Lingchang Li
Zhiming Deng
Xi Zou
Fei Li
Junhu Ji
Junshan Lei
Yangyang Liu
Hua Xiao
Huiyong Huang
Guanghui Rao
Jing Lei
Haiyan Jia
Lei Jiang

Preface

The Yangtze River, respected as the mother river of the Chinese nation, has contributed immensely toward the socio-economic development of China and braced up the national strategies such as the development of the Yangtze River Economic Belt and the integrated development of the Yangtze River Delta. Whether the Yangtze River is under good stewardship has implications on not only the well-being of more than 400 million inhabitants in the Basin, but also in broader sense the overall sustainability of socio-economic development of the whole country.

Since the founding of People's Republic of China, especially since the reform and opening-up policy was adopted, a number of skeleton water projects have been completed in succession, e.g., the Three Gorges Project, the South-to-North Water Transfer Project and the mainstream embankment strengthening. The situation of flood control and disaster mitigation and water resources utilization has been radically improved, the eco-environmental protection gradually strengthened, and the integrated river basin management constantly enhanced. All these have worked to ensure the securities of flood control, water supply, ecology and energy, leading to sustainable socio-economic development.

When inspecting the Yangtze River, General Secretary Xi Jinping eagerly urged that "we must consider the long-term interests of the Chinese nation and give our first priority to the eco-environmental restoration in the Yangtze River Basin to promote well-coordinated eco-environmental conservation and avoid excessive development," which provides the fundamental guidelines for rehabilitation and protection of the Yangtze River in the New Era. To answer the calls for well-coordinated environmental conservation of the Yangtze River, the Changjiang Water Resources Commission (CWRC), together with the institutions dispatched by line ministries and committees (Agencies) in the Yangtze River Basin, as well as related universities, research institutions and enterprises, jointly launched the League of Science and Technology Innovation for Changjiang Rehabilitation and Protection. To summarize the achievements and experience of the Yangtze River rehabilitation and protection, analyze the new situation and requirements and discuss the main issues and their solution alternatives for the Yangtze River rehabilitation and protection efforts, the Report on Yangtze River Rehabilitation and Protection 2019 (hereinafter referred to as the Report 2019) was hereby prepared. The Report 2019 was prepared by the CWRC, and the preparation of subsequent annual report will be taken over by the League.

Part I of the Report 2019 reviews the history and achievements of the Yangtze River rehabilitation and protection in the recent 70 years, identifies the new requirements for building ecological civilization and sustaining high-quality socio-economic development, proposes the general plan to build a Secure, Green, Solidary and Beautiful Yangtze River and formulates the strategic measures regarding flood control and disaster mitigation, comprehensive water resources utilization, eco-environmental protection and integrated river basin management, to ensure the sustainability of a clean and clear Yangtze River for generations to come. Part II of the Report 2019, in alignment with the characteristics of the Yangtze River Basin and the focuses of social concerns, elaborates on major topics including the Plan of Integrated Management of the Yangtze River Basin, changes of relation between the River and the Two Lakes (the Dongting Lake and Poyang Lake), inter-basin water transfer, joint operation of group reservoirs and protection of water resources and proposes major scientific and technological subjects to be studied.

Due to a tight time-schedule, a wide scope and high complexity about the Yangtze River rehabilitation and protection, plus a limited knowledge base of the authors, this Report 2019 may contain errors and inaccuracies, for which your scrutiny and corrections are very much appreciated.

Wuhan, China

Changjiang Water Resources Commission of
Ministry of Water Resources of the People's
Republic of China

Contents

Part I
Yangtze River Rehabilitation and Protection

Chapter 1
Importance of Yangtze River

The Yangtze River ranks the third in the world and the largest in China. It originates from the southwest slope of the Geladandong snow mount, the peak of the Tanggula Mountain Range of the Qinghai-Tibet Plateau, with a total mainstream length of more than 6300 km. On its way from west to east, the Yangtze River mainstream traverses 11 provinces (autonomous regions and municipalities) including Qinghai, Sichuan, Tibet, Yunnan, Chongqing, Hubei, Hunan, Jiangxi, Anhui, Jiangsu and Shanghai, flowing into the East China Sea. Its tributaries flow through other 8 provinces and autonomous regions including Guizhou, Gansu, Shaanxi, Henan, Zhejiang, Guangxi, Guangdong and Fujian. The Yangtze River drains a total area of 1.8 million km^2, representing 18.8% of the territorial area of China.

The Yangtze River is respected as the mother river of the Chinese people serving as the important underpinning for the development of the Chinese nation. With its gigantic river and lake systems, unique and complete natural ecosystems, and powerful functions of water source conservation, biological breeding, oxygen release and carbon sequestration and environment purification, the Yangtze River maintains the important biological gene treasure inventory and ecological security in China. Abundant in water, soil, forest, mineral, water energy and navigation resources, the Yangtze River ensures the national securities in terms of water supply, food and energy supply. Through rehabilitation and development, the Yangtze River brings up a population of 459 million and nurtures the splendid Yangtze River civilization, playing a prominent role in the socio-economic development.

1.1 Shelter of Ecology

Possessing an integrated system comprising mountains, rivers, forests, fields, lakes and grass, the Yangtze River Basin serves as an important biological gene treasure house. The Basin crosses tropical, subtropical and warm temperate zones,

© Changjiang Press (Wuhan) Co., Ltd. 2022

Changjiang Water Resources Commission of Ministry of Water Resources of the People's Republic of China, *Report on Yangtze River Rehabilitation and Protection 2019*, https://doi.org/10.1007/978-981-16-4927-1_1

with complex geomorphology and various ecosystems. The forest ecosystem in the western Sichuan valley, the subtropical evergreen broad-leaved forest ecosystem in the south and the wetland ecosystem on the middle and lower reaches of the Yangtze River are globally significant biodiversity protection areas and a concentrated area of rare and endangered wildlife in China, in which the wetland ecosystem in the middle-lower Yangtze River is a stopover and an important wintering site for more than 100 species and millions of trans-boundary migratory waterfowls and also a global wetland and biodiversity conservation hotspot. The forest coverage in the Basin reaches 41.3%. The areas of rivers, lakes, reservoirs, and wetlands account for about 20% of the nation's total. The Basin is abundant in a great variety of species, with the rare and endangered plants accounting for up to 39.7% and the freshwater fishes 33% of the nation's totals. There are not only rare and precious animals like Chinese sturgeon, finless porpoise, Chinese alligator, panda and golden monkey, but also rare plants like ginkgo, metasequoia and Chinese dove tree in the Basin.

The Basin, where many national-level ecology-sensitive zones are located, is an important ecological security shelter of China. Up to date, 93 national nature reserves have been set up, covering an area of 23.993 million hm^2, accounting for 30.7% and 26.3% of the national total number and the total area respectively. There are 253 national aquatic germplasm resource reserves, 255 national forest parks and 54 national geoparks within the Basin, representing 51.0%, 28.9%, 29.3% respectively of the nation's total. In addition, there are 15 world cultural and natural heritage sites and 75 national scenic areas in the Basin.

1.2 A Treasury of Resources

The Yangtze River Basin serves as the strategic tank for nationwide allocation of water resources. The Basin is relatively rich in water resources, with an average annual water volume of 995.9 billion m^3, accounting for 36% of the nation's total, ranking the first among major rivers in China. The specific water resource is 595,000 m^3/km^2, about twice as much as the national average. Every year, the Yangtze River supplies more than 200 billion m^3 of water, ensuring the water supply security for the socio-economic development in the Basin. The vast areas beyond the Basin also benefit from the development of the South-to-North Water Transfer Project, Hanjiang-to-Weihe Water Transfer Project, Yangtze-to-Huaihe Water Transfer Project, and Water Diversion to Central Yunnan, which guarantees the water supply security in these areas. In 2017, the net water volume diverted from the Yangtze River Basin reached up to 9.214 billion m^3.

The Yangtze River Basin is the major base for implementing China's energy strategy. It is the region in possession of the richest concentration of hydro-energy resources in China, with the theoretical potential of hydropower resources of 300,500 MW, and the annual energy of 2.67 trillion kWh, accounting for about 40% of the nation's total. With technically exploitable potential of 281,000 MW and the corresponding annual energy potential of 1.30 trillion kW·h, accounting respectively for 47% and 48% of the nation's total, serving as the main base of hydropower development in China. The Yangtze River Basin is also abundant in wind, solar, biological and geothermal energy, standing as the key area for new energy development in the country.

The Yangtze River is the golden waterway linking the western, central and eastern China. The Yangtze River system is rich in shipping resources, with more than 3600 navigable rivers and the corresponding total length of over 71,000 km, representing 56% of the nation's total. In the year of 2017 alone, 184 million passengers and 4.714 billion tons of cargo were transported through the waterway, accounting for 65.0% and 70.6% respectively of the nation's totals.

The Yangtze River Basin is a major grain production base in China, with a cultivated land area of 30.8 million hm^2 and a grain yield of 163 million tons, providing 32.5% of the nation's total grain production.

The Yangtze River Basin houses ample mineral resources. There are about 30 kinds of mineral resources each with reserves over 50% of the nation's totals, of which the reserves of vanadium, titanium, mercury, rubidium, cesium, phosphorus, mirabilite, and silica take up more than 80%, while that of copper, tungsten, antimony, bismuth, manganese, and thallium provide more than 50%. The iron, aluminum, sulfur, gold, silver and other mineral reserves also account for more than 30% of respective reserves of the nation's total.

1.3 Strategic Position

The Yangtze River Basin spans three major regions of southwestern, central and eastern China. According to statistics in 2017, the Basin was home to 459 million people, accounting for 33% of the country's population, and the urbanization rate reached up to 49%. The river basin is relatively densely populated, with a density about 1.8 times of the national average.

The Yangtze River Basin has formed several urban agglomerations based in the delta region, in the middle basin, in the Chengdu-Chongqing region, Hefei-centered area between the Yangtze River and the Huaihe River, in central Yunnan Province and central Guizhou Province, with more than 50 cities at or above prefecture-level. With the GDP of 29.3 trillion RMB in 2017, accounting for 35.4% of the nation's total, the Yangtze River Basin takes the position as the nation's economic focus featuring great vitality, especially the Yangtze River Delta region becomes one of the most economically flourishing areas in China.

A comprehensive traffic and transport system has been established in the Yangtze River Basin, comprising waterway, railway, highway and aviation, forming an integrated three-dimensional transport corridor in its initial stage though.

The Yangtze River is underpinning such national strategies as the development of the Yangtze River Economic Belt and the integrated development of the Yangtze River Delta. It is also a link between the Silk Road Economic Belt and the 21st Century Maritime Silk Road, integrating the opening up in inland areas and in zones neighboring the Yangtze River, sea coasts and remote areas, and possessing the unique advantage of two-way opening up eastward and westward. Therefore, the Yangtze River plays a crucial role in the nation's socio-economic development.

Chapter 2
Stages of Yangtze River Rehabilitation and Protection

The Central Committee of the Communist Party of China (CPC) and the State Council have always paid high attention to rehabilitation, development and protection of the Yangtze River. In February 1950, approved by the Central Government, the Changjiang (Yangtze River) Water Resources Commission (CWRC) was officially established, setting a new milestone for the governance of the Yangtze River by preventing and controlling floods as well as utilizing and protecting water resources in a comprehensive way. After 70 years of rehabilitation and protection efforts, the Yangtze River has turned from a flood-prone river into a river of bliss for the Chinese nation, exhibiting great vitality in driving the nation's socio-economic development.

2.1 Inception and Development (1949–1977)

Since the founding of the People's Republic of China, the Party Central Committee and the State Council have attached great importance to the flood control of the Yangtze River, which relates to the national economy and people's livelihood. Confronted with the severe situation of flood control, the CWRC studied and proposed the Report on the Program of Base Alternative for Management of the Yangtze River from 1951 to 1953, in which the three-stage management program was planned: the first stage focused on renovation and reinforcement of embankments, increasing appropriately the safe discharge on the middle and lower reaches of the Yangtze River; the second stage centered on the combined application of the embankments and the flood detention basins to store the excessive flood over the safe discharge capacity of river channels, and the third stage planned to build reservoirs in valleys to regulate floods in order to keep the water levels along the Yangtze River reaches at the safe stages. The Party's Central Committee and the State Council made deployment to draw up the Yangtze River Basin planning in the wake of floods in 1954. In accordance with the Opinions on the Three Gorges Project and the Yangtze

© Changjiang Press (Wuhan) Co., Ltd. 2022

Changjiang Water Resources Commission of Ministry of Water Resources of the People's Republic of China, *Report on Yangtze River Rehabilitation and Protection 2019*, https://doi.org/10.1007/978-981-16-4927-1_2

River Basin Planning by the Central Committee of the Communist Party of China ratified by the Political Bureau of Central Committee, CWRC prepared the Report on Key Points of Plan for Integrated Development of the Yangtze River Basin (1959), in which the first priority was given to flood control construction on the middle and lower reaches of the Yangtze River, and the Five Major Development Programs were proposed, with the Three Gorges Project as the main part together with river management, comprehensive utilization of water resources, and soil and water conservation arranged rationally. This Report on Key Points managed to coordinate the relationship between the mainstream and tributaries in integration with other aspects, and served as a guidance on water resources development on the Yangtze River Basin for a period of time, including conception of the long-term plans such as construction of the Three Gorges Project and the South-to-North Water Transfer Project, forming a grand blueprint for management of the Yangtze River. In 1972 and 1980, the Ministry of Water Resources and Power convened two symposia in terms of flood control on the middle and lower reaches of the Yangtze River, and further clarified the flood control guideline as "overall consideration of flood discharge and storage with preference on discharge". During that period, studies were made on flood control and drainage alternative solutions in plain areas of the Middle Yangtze River Basin, and master plan was completed with respect to flood control and river regulation on the middle-lower Yangtze River reaches, paving the way for the development of water resources of the Yangtze River.

In the beginning period after the founding of the People's Republic of China, the work mainly focused on engineering constructions including blocking up the historical breached embankments, overall renovating the levees on both mainstream and tributaries, and building the flood diversion works on the Jingjiang River section, with which the severe basin-wide floods in 1954 was successfully overcome. During the period of Great Leap Forward and the campaign of Learning from Dazhai Village in Agriculture Sector, a tide of water conservancy development rose, with the completion of a large number of reservoirs mainly for the purposes of water supply and irrigation and the complementary works for irrigation districts. A series of major water conservancy and hydropower projects were completed, such as the Danjiangkou, Yahekou, Bailianhe, Zhexi, Zhanghe and Lushui, and the Cut-off Works was implemented on the lower Jingjiang Reach of the Middle Yangtze River.

In January 1976, the Environmental Protection Leading Group of the State Council and the Ministry of Water Resources and Hydropower jointly approved the establishment of the Bureau of Yangtze River Basin Water Resources Protection. From then on the water resources protection in the Yangtze River Basin has been all-round unfolded.

2.2 Reform and Improvement (1978–2011)

Since the reform and opening up, China's socio-economy has developed greatly, and the rehabilitation, development and comprehensive utilization of water resources in the Yangtze River have also stepped into a new stage of rapid development. Continuous improvements have been made to the planning system. The Brief Report on Plan of Integrated Development of the Yangtze River Basin (1990) was ratified by the State Council in September 1990, and a large number of sectoral master plans and and thematic plans have been commenced, including the master plan of flood control, the master plan of integrated management of water resources, and the master plan of water resources protection, master plan of soil and water conservation, all at the Yangtze River Basin level, as well as the master plan of river regulation on the middle and lower reaches of the Yangtze River, the master plan of integrated rehabilitation and development planning of the Yangtze River estuary, master plan of the South-to-North Water Transfer and the short-term master plan of integrated rehabilitation of the Dongting Lake.

After the symposium on flood control along the middle and lower reaches of the Yangtze River in 1980, more efforts were made on construction of the key embankments including the Jingjiang Levee, Nanxian Levee, Wuhan City Embankment, Wuwei Levee, Huangguang Embankment, Tongma Levee and embankments for important polders in Dongting Lake and Poyang Lake areas. After the big floods in 1998, following the strategic arrangement made by the Party Central Committee and the State Council for reforestation in fenced-up mountains and transforming slope farming lands to forests, removing river-encroaching polders to give way back to floods and returning reclaimed lands to lakes, providing employment as a form of relief and resettling local residents to restore room for floods, reinforcing mainstream embankments and dredging rivers and lakes, the levees, on the middle and lower reaches of the Yangtze River, along the Hanjiang River as well as along the Ganjiang River, were all upgraded up to standard. Other measures were at the same time implemented, such as river regulation on the middle and lower reaches of the Yangtze River, the construction of flood storage and detention basins of 10 billion m^3 in vicinity of Chenglingji, the rehabilitation of Dongting Lake and Poyang Lake, the construction of embankments in major tributaries, and the measures of removing river-encroaching polders to give room back to floods and returning reclaimed lands to lakes were implemented, and the rehabilitation and reinforcement of reservoirs at risk, restoration of small and medium-sized rivers, and prevention and control of flash floods were carried out. With all these efforts, the flood control capability was improved significantly.

The Three Gorges Project, deemed as the backbone in the flood control system of the Yangtze River, commenced construction in 1994, and impounded water to its normal pool water level of 175 m for the first time in 2010, improving substantially the flood control situation on the middle and lower reaches of the Yangtze River and bringing immense comprehensive benefits of power generation, navigation and water supply. The construction of the South-to-North Water Transfer Project started in 2002, and the Phase I projects of both the East Route and the Middle Route were commissioned successively in 2013 and 2014.

The construction of a number of key projects was sped up and committed to operation subsequently, such as the Gezhouba, Wuqiangxi, Geheyan, Fengtan, Wujiangdu, Ertan, Pubugou, Zipingpu, Goupitan, Shuibuya, Jiangya and Zaoshi, which is significantly beneficial both to people's livelihood and the socio-economic development. Through waterway regulation and hydro complex construction, the navigation condition of the Yangtze River was greatly improved. The navigation capacity of the main channels of the Yangtze River was improved obviously thanks to the impoundment of the Three Gorges Reservoir, which drove the leap-forward development of navigation in the upstream region.

In view of the severe issues of water pollution, ecological degradation and water and soil erosion as a consequence of rapid economic development and increasing population in the River Basin, the ecological environment protection has been gradually strengthened. Actions have been taken as that programmed in the master plans of water pollution prevention and control in the Three Gorges Reservoir, in the Danjiangkou Reservoir and its upstream areas and in the middle and lower Yangtze River, as well as returning reclaimed lands to lakes and wetlands, resulting in the setup of the system of water resources protection and water pollution prevention and control based on water function zoning management. There are many records of first-time attempts, including the master plan of water resources protection at basin level, water function zoning as the base for master plan of water resources protection and management in the River Basin, environmental impact assessment on water projects, and the master plan of water quality monitoring stations and the related technical standards. Several national nature reserves were established, such as rare and unique fishes in the upstream of the Yangtze River, East Dongting Lake, Poyang Lake and Chongming Dongtan. The major national project of water and soil conservation, focusing on the upstream regions of the Yangtze River, was implemented in 208 counties (cities and districts) of 10 provinces (municipalities). The protection of natural forest resources, the construction of protective forest system, the measures of returning reclaimed lands to forests and grasses, and the comprehensive rehabilitation of water and soil erosion were put into practice. As a result, a network of water and soil erosion monitoring system was developed.

The enactment of Water Law of the People's Republic of China in 1988 got the water management on the track of rule of law. Afterwards, publication of a series of laws and regulations such as Flood Control Law, Water and Soil Conservation Law, Regulations for Resettlement of the Yangtze River Three Gorges Project, Regulations for Sand Mining Management in the Yangtze River, and Regulations for Drought Relief, provided strong legal supports and safeguards for the rehabilitation and protection of the Yangtze River.

2.3 High Quality Development and Protection (2012 to Present)

Since the 18th National Congress of the CPC, the socio-economic development in China has shifted from high-speed growth to quality growth. The Party's Central Committee vowed to speed up the progress of ecological civilization development and proposed the guidelines for water management in the New Era quoted as prioritize water conservation, balance development with spatial distribution of water resources, practice systematic governance and achieve government-market synergy. Meanwhile, a crucial strategic plan was made to advance the development of the Yangtze Economic Belt, emphasizing on promoting well-coordinated environmental conservation and avoiding excessive development.

The Master Plan of the Yangtze River Basin (2012–2030) was approved by the State Council. Such master plans were also issued successively as Master Plan of Water Resources for the Development of the Yangtze Economic Belt, Plan for Riparian Water Intakes, Effluent Outlets and Emergency Water Sources in the Yangtze Economic Belt, Plan for Eco-environmental Protection in the Yangtze Economic Belt, and Master Plan for Protection, Development and Utilization of the Yangtze River Shoreline. The comprehensive planning was carried out for major tributaries (lakes).

A number of backbone hydro projects were put into operation, such as the Xiluodu, Xiangjiaba, Jinping Cascade-I, and Tingzikou. Efforts were made to implement the river regulation on the middle and lower reaches of the Yangtze River mainstream, the comprehensive rehabilitation of Dongting Lake and Poyang Lake, and the construction of the flood storage and detention basins, as well as the rehabilitation of small and medium-sized rivers, the prevention and control of flash flood disasters, the treatment of major waterlogging-prone areas, and the reinforcement of small-sized reservoirs at risk. The construction of the flood control and drought relief system has been sped up and the hydrological monitoring station system has been improved.

Works were carried out to develop the water diversion from the Hanjiang River to the Weihe River, from the Yangtze River to the Huahei River, and diversion to the central Yunnan Province, and water resources allocation in northern Hubei Province, as well as Lijiayan Reservoir in Sichuan Province and Jiayan Reservoir in Guizhou Province for the purpose of water conservation and supply. Actions have been taken to strengthen the construction of key water source projects in the five provinces in southwest China and to initialize the complementary works and water-saving transformation in large and medium-sized irrigation districts. Efforts were made on development of waterways on the mainstream and tributaries of the Yangtze River, and Phase III of the Yangtze River Estuary Deep-water Channel Regulation Project was completed, extending the 12.5 m deep-water channel up to Nanjing City.

The protection of water resources and water ecology has been constantly enhanced. Endeavors were made on special inspections of water intakes and effluent discharge outlets, determination of ecological water demand, supervision on the minimum release from reservoirs, and the pilot work on confirmation of the property right of water flows in the Danjiangkou Reservoir area. The River (lake) Chief

System has been put into practice. Actions have been taken, in the form of Special Campaigns, to rectify indiscriminate occupation, unauthorized excavation, inappropriate dump, and unlawful construction, together with the Special Inspection of Yangtze Riverbank Protection and Utilization, and Special Rectification of Illegal Sand Mining. It has been a continuing effort to enforce the master plans of water pollution prevention and control in key reservoirs areas such as the Three Gorges Project and the Danjiangkou Project, implement natural forest protection, develop the Yangtze River riparian protective forests, and rehabilitate priority areas of water and soil erosion. New efforts are being made to ensure the standard-compliance of the protection and security facilities of the water sources for centralized water supply to main cities and towns, optimize the layout of effluent outlets to rivers in major cities and adjust them correspondingly, enhance the management and development activities within environmentally sensitive areas like nature reserves etc., enforce the fishing ban, build fish passage facilities and practice artificial fish breeding and releasing. The list of undertakings also include conducting pilot assessment on health of rivers and lakes, investigation and evaluation on ecological water demand of major rivers and lakes, assessment on the standard-compliance of the safeguard measures for major water sources, experimental ecologically-friendly operation of the Three Gorges Reservoir, and the demonstration on ecological measures to control water and soil erosion and diffuse pollution in the Danjiangkou Reservoir Area. Researches have been deepened on integrated operation of key water resources and hydropower projects, leading to inclusion of 40 reservoirs in the upper and middle Yangtze River Basin in the system of joint operation as a group.

Since 2019, the reform has been further deepened in water sector development. The underlying principles for water sector reform and development at present and in the foreseeable future will remain as "make up the short boards of water works, and reinforce supervision of the water sector". The CWRC, conforming to requirements of the times, under the premise of protection, conceived the development strategy of river rehabilitation and management in the New Era, to promote coordinated development of a secure, green, solidary, and beautiful Yangtze River.

Chapter 3
Achievements in Rehabilitation and Protection

Since the founding of the People's Republic of China, the Yangtze River rehabilitation and protection has drawn great attention from the Party Central Committee and the State Council. Under their leadership and with the guidance of plans, remarkable achievements have been witnessed in providing strong underpinning and safeguards for the sustainable socio-economic development and people's well-being in the Yangtze River Basin, by emphasizing on flood control, promoting the benefit while mitigating the damage, and supporting both development and protection with unremitting endeavors of generations.

3.1 System of Flood Control and Disaster Mitigation

The floods in the Yangtze River Basin are mainly caused by rainstorms. Determined by the distribution and coverage of the rainstorms, there are generally two typologies of disastrous floods in the Yangtze River: one is the regional flood, such as the floods in the years of 1860, 1870, 1935, 1981, 1991, 2016 and 2017; the other is the basin-wide flood, such as the floods in the years of 1788, 1849, 1931, 1954 and 1998. In addition, flash flood can emerge in hilly regions as a consequence of torrential rain that occurs in short duration and small space. The outburst flood is prone to happen as a result of dam breaches in dammed lake formed by earthquakes and landslides, and in glaciers in high altitude areas in the upstream, while the Yangtze River Delta is threatened by storm surges.

The middle and lower reaches of the Yangtze River Basin are the areas most prone to flood. According to historical records, during the 2096 years from the Han Dynasty to the Qing Dynasty, 214 events of relatively severe floods occurred, approximately once every 10 years on average. Flood disasters occur more frequently in recent years, about once every 6 years. The regions along both riverbanks of the middle and lower Yangtze River are the focus of flood prevention and control efforts, owing

© Changjiang Press (Wuhan) Co., Ltd. 2022

Changjiang Water Resources Commission of Ministry of Water Resources of the People's Republic of China, *Report on Yangtze River Rehabilitation and Protection 2019*,
https://doi.org/10.1007/978-981-16-4927-1_3

to the facts that these areas are economically developed and densely populated, and the ground elevation of their plain areas is generally several to dozens of meters below the water level in flood season, the breach of embankments means a long duration of flooding and huge losses. The flood disasters in 1931 and 1935 brought about casualties of 145,000 and 142,000 people, respectively. The flood in 1954 was the most severe one in the Yangtze River Basin over the past century, inundating 3.17 million hm^2 of farmlands and causing casualties of more than 30,000 people, making the Beijing-Guangzhou railway out of service for 100 days.

Flood control is the primary task for rehabilitation and protection of the Yangtze River. Based on the flood control policy of "overall consideration of flood discharge and storage with preference on discharge" and following the guiding principles of "win–win for both rivers and lakes, balancing between both banks of the river, and coordinating among upstream, midstream and downstream", an integrated system of flood control and disaster mitigation for the Yangtze River has basically been put in place, significantly improving the flood control capacity. The system, taking the river embankments as the basic component and the Three Gorges Project as the backbone project, features a sound coordination of mainstream and tributary reservoirs, flood storage and detention basins and river regulation works. This combines non-engineering and engineering measures, such as removing river-encroaching polders to give room back to floods, returning reclaimed lands to lakes, and water and soil conservation. The total length of the Yangtze River embankments aggregates 64,000 km, of which more than 3900 km is along the middle and lower mainstream. So far, the embankments have reached the standard-compliance figure as planned along middle and lower main reaches and along the major tributaries (for example, Yao Dike along the Hanjiang River and Ganfu Dike) and the flood control capacities of major embankments in Dongting Lake and Poyang Lake Areas, together with those along the tributaries of the Hanjiang River, the four tributaries (Xiang River, Zi River, Ruan River, and Li River) of Dongting Lake and the five tributaries (Gan River, Fu River, Xin River, Rao River, and Xiu River) of Poyang Lake have been improved greatly. 55 key reservoirs including the Three Gorges Reservoir and Danjiangkou Reservoir have been built on the mainstream and tributaries of the Yangtze River, providing a total flood control capacity of 62.5 billion m^3, and thus playing an indispensable role in retaining and regulating flood and reducing peak flow. This relieves the flood control pressure on both middle and lower reaches. The construction of 42 flood detention basins planned on the middle-lower Yangtze River has been gradually advanced. The reinforcement of 33 dikes, as well as the construction of 5 flood-diversion sluices and 4 security works have been completed, with some flood detention basins under construction in areas of Qianlianghu Lake, Gongshuangcha, Eastern Datong Lake, and Eastern part of Hong Lake. Efforts have been made to implement river regulation on the middle and lower reaches of the Yangtze River, targeted at the bank collapse sections that directly endanger the security of major embankments and some sections where the river regime changes dramatically. Actions also have been taken to remove river-encroaching polders to give room back to floods and return reclaimed lands to lakes, specifically for the middle and lower mainstream of the Yangtze River, the floodplain in Dongting Lake and Poyang Lake

Areas and 1400-plus residential polders, restoring about 16 billion m^3 of storage regulation capacity. Improvements have reached for some areas of weakness in flood control, such as small and medium-sized rivers regulation, the prevention and control of mountain torrents and severe waterlogging-prone areas, with the reinforcement works for 1029 large and medium-sized reservoirs and numerous small-sized reservoirs at risk. Non-engineering measures have been strengthened notably for flood control, such as the emergency plan for flood control and drought relief, the monitoring and warning mechanisms for mountain torrents, the flood information collection and the flood control commanding. The key reservoirs on the upper and middle reaches of the Yangtze River have been integrated into the joint operation system, which further improved the capacity in flood control and regulation.

With the continuous improvement of the Yangtze River flood control system, especially with the operation of the Three Gorges Project and the key reservoirs in the upstream, the capacity has been greatly upgraded in flood control for the Yangtze River. So far, the capacities in flood control for the main reaches of the mainstream and tributaries on the Yangtze River have roughly been up to the following standards: in case of the one-in-100-year flood on Jingjiang Reaches, the water level of Shashi City can be regulated to be no more than 44.5 m without operation of the Jingjiang Flood Detention Basin; in case of the one-in-1000-year flood or the severe flood equivalent to that in the year of 1870, the water level of Shashi City can be controlled to be lower than 45.0 m with operation of the Jingjiang Flood Detention Basins to ensure the flood running through Jingjiang reach in security level. With joint operation of the Three Gorges Project and group reservoirs on the mainstream and tributaries, complementation with the operation of flood retention and detention basins, reaches in Chenglingji and its downstream can defend the flood equivalent to that in the year of 1954. In the event of recurrence of floods equivalent to those in the years of 1931, 1935, 1954 equivalent floods, with regulation of the Three Gorges Project, the flood diversion volume and the inundated area would be substantially reduced. Thanks to the comprehensive measures, the middle and lower reaches of the Hanjiang River can defend the flood equivalent to that in the year of 1935, corresponding approximately to a return period of 100 years. The Ganjiang River can defend floods with return periods of 20–50 years, and the other tributaries can defend floods with return period of 10–20 years.

3.2 System of Integrated Management of Water Resources

The Yangtze River Basin is abundant in rainfall, with an average annual precipitation of about 1100 mm, but local droughts occur from time to time, as a result of uneven temporal-spatial distribution, coverage of a wide arrange of longitude and latitude in the Basin with characteristics of striking disparities in topography, climate and soil conditions. The upper Yangtze River is vulnerable to frequent spring and summer droughts, therefore prone to continuous droughts from winter to spring and from summer to autumn. In the middle and lower Basin, the summer and autumn droughts dominate, with frequent droughts in summer or from summer to autumn. Most human activities in history were of agriculture, and the drought disasters recorded in history usually referred to agricultural losses. For example, the severe drought in 1934 affected a vast area of 6.67 million hm^2 in Jiangsu, Anhui, Hunan, Hubei, and Jiangxi provinces, leading to a yield reduction of 80%. With the socio-economic development, the increase of population and the acceleration of urbanization, the drought and water shortage have gradually extended from agriculture to animal husbandry, urban and rural water supply and rural drinking water. The Yangtze River Basin suffered from severe droughts in the dry years of 1972, 1978, 2006 and 2007. From 2009 to 2010, five provinces (autonomous regions, municipalities) located in southwest China experienced consecutive droughts. Such areas are the major ones affected by drought disasters as the Central Yunnan Plateau, Central Guizhou Region, Hengshao Hill, Southern Hunan Region, Jitai Basin, Nanyang Basin and Humpy Ground in Northern Hubei. Some other areas in the Basin also have, though to various degrees, the water shortage problems due to lack of water resources, water infrastructures, and/or as a result of water quality degradation.

At present the system of the integrated management of water resources in the Yangtze River Basin has taken shape. The water resources allocation system has been basically completed with large and medium-sized key reservoirs, water diversion works, pumping projects and water transfer projects as main components, improving the water supply security all round. The total number of reservoirs in small, medium and large size built in the Basin has aggregated to 51,900, with a total storage capacity of 414.1 billion m^3. The number of water diversion projects has cumulated to 368,000, providing 206 billion m^3 of water for the whole river basin in 2017, which plays an effective part in satisfying the water demand for the socio-economic development in the Basin. The security has been attained in rural drinking water, and is constantly improved in water supply security in cities and towns. The foundation work of irrigation and drainage infrastructure in rural areas has been continuously enhanced, with completion of 156,000 irrigation districts and 16.3 million hm^2 of effectively irrigation district, of which the water-efficient irrigation district has aggregated to 1.57 million hm^2. Positive results have been achieved in building a water-saving society, as showing in the data of the Yangtze River Basin in 2017: the reuse rate of industrial water was about 75%, the leakage rate of urban water supply network was nearly 18%, and the water utilization coefficient of irrigation was approximately 0.52. 25 cities in 8 provinces (municipalities) including Sichuan, Chongqing, Guizhou, Hubei, Hunan, Jiangxi, Anhui and Jiangsu were chosen as the pilot sites to build a water-saving society. The inter-basin water transfer figuration is gradually taking shape, with the role of Yangtze River as a strategic water source becoming increasingly apparent. The Phase I projects of both Middle and East Routes of the South-to-North Water Transfer Project have realized the goal of transferring water, with a total of 22 billion m^3 of water transferred to Beijing, Tianjin, Hebei, Henan, Jiangsu and Shandong by 2018. Such water transfer and diversion projects have been started successively as Hanjiang-to-Weihe Water

Diversion Project, Yangtze-to-Huaihe Water Diversion Project, Water Diversion to Central Yunnan, and the water resources allocation in northern Hubei Province, with these efforts, a macro crisscrossing network of water resources allocation was formed, featuring water transfer from south to north and water reallocation between east and west.

Remarkable progress has been achieved in the hydropower development in the River Basin. A series of large hydropower projects have been completed in succession, such as Gezhouba, Three Gorges, Xiluodu, Xiangjiaba, with Wudongde and Baihetan stations under construction. In 2017, the total installed capacity of hydropower stations (constructed and under construction) in the River Basin reached 237,000 MW, accounting for 74% of the nation's total. The shipping industry on the Yangtze River has been in a booming period, with shipping conditions greatly improved. According to statistics in 2017, the navigable mileage on both mainstream and tributaries was 71,000 km and the freight volume aggregated 4.714 billion tons.

3.3 System of Eco-environmental Protection

The water resources of the Yangtze River Basin are generally in good quality but face the striking contradiction between socio-economic development and eco-environmental protection. The total amount of waste effluent is very huge, 35.32 billion tons in 2016. What is worse, the aquatic biodiversity indicators of the Yangtze River are continuously declining, and many rare species are at the risk of extinction. According to the results derived from the water and soil conservation census included in the first nation-wide water resources census, by the end of 2011, the total area of water and soil erosion in the Yangtze River Basin was 384,600 km^2, of which the hydraulic erosion was 361,100 km^2 while the wind erosion accounted as 23,500 km^2.

Based on water function zoning management, the system of water resources protection and pollution prevention and control has been initially set up, continuously strengthening the point, non-point source and endogenous pollution treatment. Efforts have been made to implement the plans for water pollution prevention and control in areas including the Three Gorges Reservoir and its upstream region, as well as the middle and lower reaches of the Yangtze River. Facilities for collecting and disposing urban and rural sewage have been built with vigorous efforts. The industrial pollution sources are treated in a more forceful way. A mix of measures have been carried out energetically, targeting at treatment of black and odorous water body in urban and rural areas, prevention and control of non-point pollution from agricultural activities, treatment of livestock breeding pollution, clearance of seine breeding and removal of garbage along rivers and lakes. The emergency plans were launched for sudden water pollution incidents are prepared in time. At present, the water quality of the Yangtze River is generally premium, and most of the water bodies in the Basin can meet the water quality requirements by water use in their respective water function zoning. According to statistics concerning the Yangtze River in

2017, the length of river where its water is classified as Grade I-III accounted for 83.9% of the evaluated river length. The number of the cross-sections with water quality classified as Grade I to III occupied 89.6% among 164 evaluated provincial boundary cross-sections. Amongst 61 lakes and 362 reservoirs, the waters with the quality reaching Grade I-III took up 14.8% and 81.8% respectively; amidst 515 evaluated water sources, 73.2% were up to the standard all year round, and the ones with standard-compliant rate of above 80% accounted for 90.3%.

Actions have been taken to establish the national-level nature reserves including the upper reaches of the Yangtze River for unique rare and precious fish, the East Dongting Lake, the Poyang Lake, and the Dongtan Reserve at Chongming of Shanghai. Measures to protect fish species also include aquatic vegetation restoration, protection by resorting to alternative habitats, conservation and restoration of fish's spawning, feeding and wintering grounds, restoration of river connectivity and artificial fish nest construction, artificial fish breeding and releasing, and returning reclaimed lands to lakes and wetlands. All these efforts elevate the protection for rare and unique fish, and enhance the protection and restoration for wetland nature reserves on the middle and lower reaches of the Yangtze River and for international major wetlands. Ecological regulation experiments for water projects have been carried out to promote the natural reproduction of the four major carps in the mainstream Yangtze River. The restoration and management of the water ecological environment have effectively curbed the functional degradation of water and wetland ecology.

Measures have been taken to protect natural forests, develop the protective forests system along the Yangtze River, return the reclaimed lands to forests and grasses, and comprehensively rehabilitate the soil eroded lands in continuous patches. The afforestation area has reached 5,049,700 hm^2, the area of forests converted from farmland has accumulated to 5,727,900 hm^2, and the stony desertification treatment has covered an area of 3,573,300 hm^2. The area of water and soil erosion in the River Basin experienced a historic transformation from increasing to decreasing in 2013. By 2017, the water and soil erosion treatment had covered a total area of more than 430,000 km^2. It has been witnessed that the ecological environment in the regions under treatment has been substantially improved.

3.4 System of Integrated River Basin Management

Continuous progress has been made in the legal construction concerning the Yangtze River Basin. A series of laws, regulations, and administrative rules have been successively promulgated and enforced, including Water Law, Flood Control Law, Regulations for Sand Mining Management in the Yangtze River, and Administration Procedures for Operation of the Three Gorges Reservoir and for the Management of Water Resources and River Channels in the Reservoir Areas. Progress has been achieved in perfecting the water resources management system, which integrates the river basin management with regional management, leading to an enhanced level in integrated management and water resources regulation. The river basin planning system is becoming increasingly mature, composed of integrated plan and master plan, which complement each other, for the Yangtze River and its major tributaries (lakes), playing a leading and binding role in planning. Water administrative approval and management have been consolidated on water withdrawing permits, flood impact assessment, water and soil conservation schemes for production and construction projects, etc. Coordination and consultation mechanisms have been gradually established for flood control and drought relief, water and soil conservation, sand mining, water resources protection and water pollution control, water volume arrangement, and joint operation of group reservoirs. Unprecedented efforts have been made to protect river and lake shorelines, with integrated law enforcement inspections along rivers and lakes conducted for construction projects, aiming at sternly investigating and cracking down on cases of illegal occupation of shorelines and illegal sand mining in accordance with law. Inter-provincial water disputes have been positively dealt with in such water bodies as Qingyi River, Shuiyang River, Chuhe River, Dongting Lake, and Danjiang River. With all these efforts, water-related affairs have been maintained in a good order in the River Basin.

Capabilities have been improved significantly in providing scientific and technological support. CWRC has continued carrying out hydrologic and geological surveys and investigations, and thus accumulated a large amount of data on hydrology, topography, geology and engineering. Informatization has been accelerated. Specifically, having a hydrological station network well established, CWRC has preliminarily set up the system of monitoring information collection and real-time surveillance, and initiated the top-level design of the Smart Yangtze River, strengthening the integration of Programming Water Management by One Diagram with the river governance cause. Strong scientific and technological capacities have been accumulated, manifested by establishment of 11 national and provincial (ministerial) engineering and technology research centers and key laboratories, numerous awards at national and provincial (ministerial) level due to more than 600 major achievements and a number of world-leading core technologies. So far, CWRC has extended its international cooperation and exchanges to more than 60 countries and regions, and gathered a great many talents specialized in hydrology, survey, planning, design, scientific research and management.

Chapter 4
Situation Facing the Yangtze River Rehabilitation and Protection in the New Era

4.1 Main Problems

Although remarkable achievements have been made in integrated management of the Yangtze River, the demands of socio-economic development and the call for well-coordinated protection of the Yangtze River in the New Era are manifesting the emerging challenges resulting from unbalanced and insufficient development of water sector, which are expounded as the following.

4.1.1 Inadequacy in Flood Control and Disaster Mitigation System

On the middle-lower reaches of the Yangtze River, the prominent problem is that the floods with high peak level and huge volume still overwhelm the discharge capacity of the river channel and the storage of the lakes. In case of the flood equivalent to that in the year of 1954 in the middle and lower reaches, it is by all means imperative to divert a large amount of flood waters off stream according to the planned flood defense standard. However, the construction of flood storage and detention basins lags far behind, which means the detention basins haven't met the desired scenario of flood being able to be diverted into, be retained and detained in and be drained out whenever necessary. After completion of the upstream key reservoirs, less sediment-laden water will be released downstream of the Yangtze River for longer, which will result in potential bank collapses and river regime changes on some sections, threatening the flood control security. Changes in the river–lake relation also affect their storage and discharge capacities. At present, the flood control capacities of embankments are relatively poor along some upstream trunk sections, major tributaries and lakes, and some tributaries are in urgent need of building flood control reservoirs or tapping such potentials from existed reservoirs. Efforts have to be intensified to push forward the

© Changjiang Press (Wuhan) Co., Ltd. 2022

Changjiang Water Resources Commission of Ministry of Water Resources of the People's Republic of China, *Report on Yangtze River Rehabilitation and Protection 2019*,
https://doi.org/10.1007/978-981-16-4927-1_4

regulation of small and medium-sized rivers, the prevention and control of mountain torrents, and the enhancement of reservoirs at risk. Floods can hardly be drained out in some key areas because of an insufficient pumping capacity. Non-engineering systems need to be improved, such as laws and regulations concerning flood control, organization and commanding, emergency management, social security and technical support.

4.1.2 Weakness in Development and Conservation of Water Resources

The water resources have the characteristics of uneven spatial and temporal-spatial distribution in the River Basin, which does not match the demands of the socio-economic development. Such arid regions are short of water resources and incompetent in regulation with insufficient regulation capacities as the Central Yunnan Plateau, Central Guizhou Region, and Hengyang-Shaoyang hilly regions, featuring sharp conflict between water demand and supply. With the unbalanced development of water supply and irrigation infrastructure among regions, the task is arduous in building a water-saving society, leaving problems of low efficiency and extensive pattern of water utilization essentially unsolved. Awareness is still weak in water-saving irrigation in agriculture and the driving force is insufficient in sustainable development. What's more, the efficiency is not high in water and fertilizer utilization, and the agricultural non-point source pollution is severe from agricultural activities. All these put water saving and emission reduction, as well as efficient water-saving irrigation, in a serious situation. Areas of weakness still exist in the construction of emergency water sources in some cities and the safety for rural drinking-water still needs consolidation and improvement. During dry years and dry seasons, it is necessary to coordinate the conflicts in water uses among some tributaries, among domestic, productive and ecological purposes, and between host area and borrow area in inter-basin water transfers. The multi-objective joint operation system needs to be further enhanced for group reservoirs on the mainstream and tributaries in the upper Yangtze River. The lower reaches of the Yangtze River have a great capacity to divert water, however, the potential issues on water use security cannot be ignored. The mainstream waterway on the middle reaches of the Yangtze River still has local problems i.e. so called "intestinal obstruction", which has blocked the full play to its role as the golden waterway.

4.1.3 Arduous Tasks in Eco-environment Protection and Restoration

Issues such as water pollution, water ecological damage and water and soil erosion are still obvious in some parts of the River Basin. The water pollution phenomenon can be seen in some major tributaries like Minjiang River, Tuojiang River and Xiangjiang River. Polluted belts appear in the coastal waters in some major riparian cities such as Shanghai, Nanjing, Wuhan, Chongqing and Panzhihua. Eutrophication exists to a medium degree in Dianchi Lake, Chaohu Lake and Taihu Lake. Some centralized drinking-water sources in the River Basin do not always provide acceptable water quality throughout the year. There are numerous potential risk sources of water pollution and high risks of water pollution bursts. The degraded natural connectivity between rivers and lakes and the changes in habitat conditions pose a threat to the diversity and integrity of aquatic organisms, manifested by the shrinking wetlands with degenerative major functions and changes in the habitat environment such as the disappearance of some natural spawning grounds. Up to now, there are still more than 300,000 km^2 of water and soil erosion area and more than 6.67 million hm^2 of slope cultivated land in the River Basin, remained to be effectively treated. The water and soil erosion induced by some production and construction projects is still a rather difficult problem. The development along part of the shorelines severely conflicts with the water ecological environment protection in some river sections. Due to chaotic and predatory exploitation of hydropower resources in some rural areas, as well as shortage of ecological discharge facilities, it is common for some rivers to suffer from the flow reduction and or even flow cutoff in some river reaches.

4.1.4 Urgent Call for Integrated River Basin Management

Laws and regulations of river basin management needs improvement, inter-sectoral and inter-regional coordination is insufficient, the dispute settlement mechanism is not perfect, and the system of payment for ecological services of river basin is establishing but in slow progress. There is much room for improvement of inter-sectoral cooperation in sand mining, shoreline management and water project operation. The information monitoring still has a way to go to cover the whole basin, informatization level needs improvement, and channels for information exchange and sharing need to be cleared, which indicates an urgent need for more mature informatization. There is a gap in the comprehensive law enforcement capability, demonstrated by a shortage in rigid constraint, implementation and supervision of the integrated river basin plan, as well as the incompleteness of the unified monitoring system. No joint force has been formed in terms of studies and researches on river basin rehabilitation and protection, therefore more efforts are needed to make breakthroughs on major issues, and science and technology are needed more as a support.

4.2 New Circumstances

The 19th National Congress of the Communist Party of China has specified the ambitious goal of securing a decisive victory in building a moderately prosperous society in all respects and embarking on a new journey to fully build a modern socialist country. Against the backdrop that China has ushered into a new stage of ecological civilization and quality development both in economy and in society, people's longing is getting more urgent for water security, quality water resources, healthy water ecology, and pleasant water environment. At the new historical starting point, the cause of rehabilitation and protection of the Yangtze River has come to embrace crucial development opportunities, as well as meet new requirements.

4.2.1 Requirements on Building an Ecological Civilization

Building an ecological civilization is vital to sustain the Chinese nation's sustainable development. The Party Central Committee and the State Council have attached great importance to building ecological civilization, and thus put forward the goal of speeding up the construction of ecological civilization and building a beautiful China. Since the 18th National Congress, the Party Central Committee has shown great foresight in proposing the "five-sphere integrated plan", of which the ecological civilization construction is an important part. In 2015, the Party Central Committee and the State Council issued the Opinions on Accelerating the Construction of Ecological Civilization, placing building an ecological civilization in a prominent strategic position. General Secretary Xi Jinping always highlights the rehabilitation and protection of the Yangtze River, and he stressed at the symposium on promoting the development of the Yangtze Economic Belt that "we must consider the long-term interests of the Chinese nation and give our first priority to the restoration of the Yangtze ecological environment to work together for promoting well-coordinated environmental conservation and avoiding excessive development".

Driven by the requirements for building an ecological civilization, we need to further strengthen resources conservation and protection. In terms of water resources utilization, we must reinforce the demand management on water resources utilization, optimize water resources allocation, and emphasize water conservation to better build a water-saving society. We need to penetrate the conception concepts of green, cycle, and low-carbon development into the Yangtze River rehabilitation and protection. In detail, human activities must not break the upper limit of resource consumption, the bottom line of environmental quality, and the red line for ecological protection, but improve the utilization efficiency of water resources, and reduce the pollutant discharge. We need to make more efforts in reinforcing the maintenance and protection of flood discharge and storage space, and better managing shorelines and sand mining, in a bid to promote the harmony between human and water. More ecological protection and restoration are required, which highlights the idea of mountains, rivers, forests, farmlands, lakes and grasslands constituting the life community and is in accordance with the principle of respecting the nature. Actions are called for strengthening the ecological restoration of major waters, speeding up the restoration of river and lake ecology and the recovery of biological channels, safeguarding the ecological water volume kept in rivers and lakes, and intensifying the treatment of water and soil erosion in basins. We also need to give full play to the ecological benefits of water conservancy projects and adopt a holistic approach to protect the natural ecosystems and environment in a more vigorous way, in order to build a beautiful Yangtze River where human and nature coexist in harmony.

4.2.2 Requirements of Socio-economic Development

The 19th National Congress of the CPC has thrown light on the striving goals and strategic arrangements by stages from now to the middle of this century, where water conservancy is a fundamental part and also an important support to build China into a great modern socialist country that is prosperous, strong, democratic, culturally advanced, harmonious and beautiful and to meet the people's needs for a better life.

Spanning the eastern, central and western parts of China, the Yangtze River Basin is the carrier of the national strategies quoted as the development of the Yangtze Economic Belt and the integrated development of the Yangtze River Delta region, as well as an important link between the Silk Road Economic Belt and 21st Century Maritime Silk Road Initiative, supporting the tomorrow of cluster cities in the Yangtze River Delta, the middle Yangtze reaches, and the agglomerations like Chengdu-Chongqing. The Outline Plan for Development of the Yangtze Economic Belt defines the strategy on developing the Yangtze Economic Belt as the pilot demonstration belt for building an ecological civilization, the innovation-driver belt for leading nationwide transformative development, the inland river belt with global influence, and the coordinated development belt for linking interaction and cooperation among eastern, central and western parts of China. In the master plan of national main function zoning, the Yangtze River Delta is a region for optimized development, where the area between the Yangtze River and Huaihe River, the middle Yangtze reaches, Chengdu-Chongqing region, the Central Guizhou region, and the Central Yunnan region are deemed as the key development regions.

As the goal of building a moderately prosperous society in all respects is being approached day by day and with the steady progress in constructing socialist modernization, the socio-economic development in the Yangtze River Basin is certain to meet with profound changes. New and higher requirements for strengthening the River Basin water safeguards are proposed since we call for promoting quality and green economic development, ensuring and improving people's well-being, implementing the coordinated regional development and building a beautiful China. In order to underpin the national and regional coordinated development strategy, it is imperative to develop quality water infrastructure to address the mismatch between the water resources distribution in the River Basin and the layout of productive forces, and to find overall solutions to water security, water resources, water environment, and water ecology. To fully guarantee flood control security, it is necessary to speed up filling areas of weakness in flood control and build a sound system of flood control and disaster mitigation that matches the current socio-economic development. Efforts has to be made to ensure urban and rural water supply security and improve the level of water conservation and efficient utilization, so as to further optimize water resources allocation in the River Basin and make up for inadequacies in integrated water resources utilization. We should do more to intensify eco-environmental protection and promote the virtuous cycle of ecosystems. It is our duty to enhance integrated river basin management and give full play to the role of the market in resources allocation, forming and basically establishing a water governance system corresponding to the current level of modernization.

4.2.3 Challenges by Physical Environment Changes

Affected by global climate changes, the local climate of the Yangtze River Basin is abnormal in recent years, with frequent occurrence of extreme weather incidents. In 2006, the upper and middle reaches of the Yangtze River experienced severe drought. At that time, the annual runoff at the Yichang hydrometric station ranked the last in the data series collected during the past 141 years from 1878 to 2018. In the early to middle November of 2008, a rare late autumn flood occurred on the upper reaches of the Yangtze River, with the peak flow at the Cuntan hydrometric station ranking the first compared with the same period since the existence of actual measured data. From 2009 to 2010, five provinces (municipalities, autonomous regions) experienced continuous droughts in extensive regions in southwest China including Yunnan, Guizhou, Sichuan, Chongqing and Guangxi. During the flood season in 2016, the maximum rainfall occurred in many places in Hubei Province since the existence of actual measured data.

Earthquakes and barrier lakes have posed serious threats to socio-economic development. The earthquakes occurring respectively in Wenchuan in 2018 and in Yushu in 2010 caused severe damage to the local mountains and a large number of water conservancy infrastructures, leading to heavy casualties and financial loss. In 2018, two successive landslides occurring along the Jinsha River led to barrier lakes. Although casualties were avoided through scientific regulation, the downstream infrastructure was still destroyed and the safe operation of the downstream reservoirs was under serious threat.

Chapter 5
Visions on Yangtze River Rehabilitation and Protection

In the New Era, the rehabilitation and protection of the Yangtze River will be advanced in accordance with the five-sphere integrated plan and the four-pronged comprehensive strategy. We must pursue the vision of innovative, coordinated, green, open, and sharing development, stay commitment to the people-centered philosophy, and fully act on the water conservancy guiding principle quoted as "prioritize water conservation, balance development with spatial distribution of water resources, practice systematic governance and achieve government-market synergy". We also should stick to the conception of giving the priority to ecology and green development, as well as endeavoring to protect the environment while pursuing development and achieve development in a well-protected environment. The water security safeguard system needs to be improved and the pleasant ecological environment requires maintenance. With all of these efforts, the sustainable development of our economy and society can be supported, and the clean water from the Yangtze River would benefit later generations.

By 2035, the safeguarding capacities will be fully improved in flood control security and rural and urban water supply security in the Yangtze River Basin, and the level of water-saving and the efficiency of water use will be significantly enhanced. By that time, a water-saving society will have taken its initial shape, where the ecological environment will show an overall improvement and a modern water infrastructure network will be completed. Such an elevated level of integrated river basin management makes it possible to formulate a modern water governance system that matches the socialist modernization in the initial stage. By the middle of this century, the target to modernization of water conservancy will be completely hit, water security will be fully guaranteed, and the capabilities will stand in the leading position in the world with respect to protecting, conserving, utilizing and managing water resources. Based on the above conditions, the vision will entirely come true in realizing a Yangtze River that is secure, green, solidary, and beautiful.

© Changjiang Press (Wuhan) Co., Ltd. 2022

Changjiang Water Resources Commission of Ministry of Water Resources of the People's Republic of China, *Report on Yangtze River Rehabilitation and Protection 2019*,
https://doi.org/10.1007/978-981-16-4927-1_5

5.1 Secure Yangtze River

The Principle should be put into practice on the Nine Adherence required by General Secretary Xi Jinping about improving the capacity in coping with natural disasters. To this end, fresh endeavors must be made to consummate the engineering system for flood control, strengthen the risk management, and establish a scientific and efficient system of non-engineering measures. This vision, with enhanced capacity in disaster control and mitigation, will allow the flood water to be confined in its own way and discharged safely, making the Yangtze River a secure river that only brings benefits to the people.

- Harmony Between Man and Water

To build an ecological civilization, we should respect, conform to and protect the nature, and coexist in harmony with it. In order to properly cope with the relationship between human and water, we should stick to the following rules: giving full play to people's subjective initiative on water management while obeying the objective laws by revering water; transforming and making use of the nature while taking the initiative to adapt to and protect the nature, giving the way to flood.

- Overall Consideration

The engineering and non-engineering measures should be combined under the flood control guideline of "overall consideration of flood discharge and storage with preference on discharge", as well as the principles of "win–win for both rivers and lakes", "balancing between both riverbanks and coordinating among upstream, midstream and downstream". We should pay equal attention to construction and management, with vision to long-term while realizing short-term programs, and highlight priorities. On the prerequisite of ensuring flood control security, we need to give full consideration to the utilization of flood resources and the protection of the water ecological environment.

- Proper Standards

Flood control works are the fundamental infrastructures to guarantee flood control security in the River Basin. With this connotation, every region should make up for the inadequacies in flood control infrastructures based on the local characteristics. The construction standards for flood control works should keep in line with the level of socio-economic development, with the financial capacity of both the state and local governments, with the after-disaster-event influence on the society and economy, and with the resilience of the reconstruction.

- Flood Management

We can give full play to the role of the flood control engineering system through scientific regulation and management. Specifically, human activities should be well regulated to protect river channels, lakes, flood storage and detention basins and floodplains from being encroached on; the public awareness on flood control should be raised; and great attention should be paid to flood risk transfer and benefit compensation. All these measures help improve the capabilities in risk tolerance and resilience in self-recovery after disasters.

5.2 Green Yangtze River

Actions are taken to push forward in depth the nation-wide water conservation campaign, improve the water allocation system for efficient utilization, strengthen the integrated water resources regulation and allocation, and promote the shift of economic development modes and people's lifestyle, so as to build a green Yangtze River characterized by efficient conservation and rational allocation of water resources.

- Priority on Water Conservation

We should tackle the problems of shortage of water resources and deterioration of the water ecological environment following an integrated and comprehensive approach. To achieve this goal, on one hand, we ought to pay more attention to demand management with the total water use quantity strictly controlled and the water use efficiency elevated, while waging a national campaign of raising the awareness of the general public on water saving in the everyday life of them. On the other hand, we should implement rigorous management on water use, and exercise water conservation throughout the whole process of water infrastructure construction and socio-economic development, in order to support the sustainable use of water resources.

- Scientific Allocation

To shape a flexible, secure and efficient allocation pattern of water resources, we should take overall consideration of domestic, productive and ecological water uses, coordinate the water demands within and beyond the River Basin as well as in the upper and lower reaches, formulate reasonable plans for river flow allocation, and optimize the regulation of local and introduced water, surface water and groundwater, as well as conventional water and reclaimed water, in order to establish a water resources allocation pattern with flexible, secure and efficient characteristics. These efforts will boost the harmonious development of water resources, socio-economy and ecological environment, and better facilitate the national water resources allocation pattern featuring water transfer from south to north and water reallocation between east and west.

- Comprehensive Utilization

Taking the socio-economic development and long-term interests into account, we should give overall consideration to the comprehensive water resources utilization covering water supply, irrigation, navigation and power generation, and coordinate the respective interests and conflicts among different sectors and industries of water conservancy, electric power, agriculture, urban construction, transportation, environmental protection etc., as well as between upstream and downstream, left bank and right bank, mainstream and tributaries in the River Basin. Only by arranging scientifically, can we realize the reasonable development and sustainable utilization of water resources.

- Strengthened Regulatory Control

We should keep development activities under control rationally, facilitate the dual control on total volume and intensity of water resources consumption, conduct evaluation on the water resources carrying capacity, and establish a corresponding monitoring and forecasting mechanism. We must do more work to adhere to the principle of determining the urban spatial layout and scale, the construction land use, the population size, and the industry scale and structure based on water resources, in order to force the industries to transform and upgrade more quickly. The socio-economic development shall be continuously promoted to correspond to the water resources carrying capacity. Strict control and management are required for the ecological flow and water volume in rivers and lakes, and integrated regulation and management need to be intensified on river basin water resources, so as to realize sustainable use of water resources.

5.3 Solidary Yangtze River

Under the unified deployment of the Party's Central Committee, we should vigorously promote the integrated river basin management that is in concordance with the times and features of the Yangtze River, which calls for a river basin management system and mechanism featuring broad-based participation, democratic consultation, joint decision-making and decentralization of responsibilities. In this connection, we need to enhance our managerial capacity, so as to build a Solidary Yangtze River with co-consultation, co-development, co-sharing and win–win mechanism.

- Sound Regulations

The governance of the country and of the water is on the track of rule of law. By centralizing social consensus on the Yangtze River rehabilitation and protection, legal bottom line will be demarcated, and specifically the Yangtze River Protection Law, being in the process of legislation, will provide a legal framework to form a consummate system of laws and regulations as a legal base for governing all the water related activities.

- Well-defined Institutional Arrangement

There is a need to further clarify the management responsibilities among river basin authority and regional jurisdictions, as well as among different line ministries and departments, so as to set up a river basin management mechanism featuring clearly defined competence, unequivocal decentralization of responsibilities, compliance with codes of conduct, and synergy in operation. The role of local governments in River (Lake) Chief System shall be underscored, and coordinated joint actions should be stressed among water related management departments.

- Unified Management

A management system will be put in place featuring unified monitoring, planning, scheduling and supervision among different regions and departments with overall consideration of the requirements for flood control and drought relief, water supply security, and ecological protection in the River Basin.

- Social Co-governance

Such work emphasizes a system where the government takes the lead and the public participate, integrating resources from all sectors and regions. Joint efforts are made to put major measures into practice for ecological restoration of sensitive waters, improvement of river–lake relations, protection of water sources, and environment rehabilitation etc., a mechanism of payment for river basin ecological service will be established to facilitate social co-governance.

- Capacity Building

We can integrate and improve the water related information monitoring stations in the River Basin and set up various monitoring indicator systems and assessment control standards, in order to enhance capacity and level in informatization management for strong basin-wide watershed supervision. Actions should be taken to strengthen studies and researches on basic theories and key issues related to river-basin rehabilitation and protection, plan the construction of research platforms and the corresponding complementary facilities, and consolidate the application of research outcomes.

5.4 Beautiful Yangtze River

Keeping a strong commitment to the ecological civilization which requires us to stick to the policy of putting conservation and protection first and mainly relying on the natural recovery of the environment, we are invigorated to fully implement eco-environmental protection and recovery, so as to build a Beautiful Yangtze River with clean water, green banks, smooth flow and pleasant landscape.

- Systematic Protection

Systematic protection requires a systems approach that integrates mainstream and tributaries and all the elements in the ecosystem-mountains, rivers, forests, farmlands, lakes and grasslands as an organic integrity. Following this approach, we can plan as whole for both land and water, urban and rural areas, rivers, lakes and oceans; coordinate upstream, midstream and downstream; integrate the aspects of water resources, water ecology, and water environment; reconcile industrial layout, resources development and eco-environmental protection; and regulate scientifically all types of water resources and hydropower projects, so as to tap comprehensive benefits of water resources and build an integrated pattern for eco-environmental protection at basin level.

- Exercising Zoning-Specific Measures

We must rely on the master plan of main function-oriented zoning, strengthen the zoning-based management of water and ecological environment, and systematically build a pattern for ecological security. In order to cope with the prominent problems in key areas and the functional orientation of ecological environment in the east, central and western regions, the upper, middle, and lower reaches, as well as mainstream and tributaries, we should practice differentiated protection strategies and management measures to implement precisely targeted rehabilitation. On the basis of prospective long-term program, we should actively promote river basin water eco-environmental protection and restoration in view of prominent problems, so as to achieve the integration between overall progress and key breakthroughs.

- Reinforcing the Bottom-Line Management

We should demarcate the upper limit for resources development, red line for ecological conservation, and bottom line for environmental quality. Efforts should be made to strengthen rigid constraints for the ecological environment, in a bid to ensure the eco-environmental quality of the Yangtze River should to get better but not worse. The negative list management system should be implemented to strictly specify the management and control measures on shorelines, river sections, regions and industries.

- Improving the Quality

Upholding the development idea of clean waters and green mountains being invaluable assets, we are encouraged to strengthen the protection and restoration of the ecological environment to ensure the smooth flow in rivers and lakes, foster a water ecological security shelter, protect the biodiversity and rare species to guarantee the health of rivers and lakes, and improve the ecosystem service functions, to faster form new spatial patterns that are resource-saving and environment-friendly.

Chapter 6
Strategic Measures

6.1 Flood Control and Disaster Mitigation

6.1.1 Reinforcing the System of Flood Control Works

Efforts should be made to strengthen the construction of embankments, give full play to the flood control-based reservoirs, promote construction of flood storage and detention basins, and rehabilitation of river channels. These measures make up for inadequacies in flood control works and put in place a system of flood control works at appropriate level.

6.1.1.1 Improving the Embankments

Embankments are the fundamental infrastructure for flood control of the Yangtze River, for which standard-compliance construction should be completed stage by stage. Embankments and bank defensive works will be built along rivers running through cities above the county level, towns with a dense population, concentrated and extensive farmlands and flat areas. The rehabilitation will be continuously strengthened for major tributaries including Hanjiang River, Jialing River and Chuhe River. Efforts will be initiated for the seepage prevention in dike foundation and the defect treatment in dike body for 11 key polders such as Lannihu Lake in Dongting Lake Area, and 20 key polders such as Hongqilianwei in Poyang Lake Area to consolidate the flood control capacity. The embankments will be heightened and reinforced for ordinary polders in the Dongting and Poyang Lake Areas. Typical sections will be selected as pilots for ecological rehabilitation on embankments along the mainstream of the Yangtze River. The construction of embankments will be completed for such large lakes as Chaohu Lake, Liangzi Lake, Honghu Lake, and Caizi Lake, etc.

© Changjiang Press (Wuhan) Co., Ltd. 2022

Changjiang Water Resources Commission of Ministry of Water Resources of the People's Republic of China, *Report on Yangtze River Rehabilitation and Protection 2019*,
https://doi.org/10.1007/978-981-16-4927-1_6

Efforts will be made to continuously strengthen the rehabilitation of small and medium-sized rivers, giving priorities to the sections where it is prone to flood and waterlogging disaster, where the protected areas are densely populated and where the targets of protection are important. Focus is placed on the reinforcement and construction of embankments, bank defensive works and dredging of river channels in maintaining the room and the capacity of flood discharge. While improving the flood control capacity, we should also strength the ecological restoration and protection in combination with the construction of cities/towns and Beautiful Villages.

6.1.1.2 Giving Full Play to the Role of Reservoirs in Flood Control

The construction of flood control-based reservoirs plays an important role in improving the capacity in flood control on the middle and lower reaches of the river where it is located. On the basis of identifying and coordinating the conformity between reservoir functions and ecological conservation red line as well as the management and control requirements, the construction should be pushed forward for planned flood control-based reservoirs. The reinforcement and defect treatment should be completed for planned small reservoirs as soon as possible, the safety of reservoirs and dams should be assessed on a regular basis and the inconvenient, if any, should be corrected in time. Investigations should be conducted on any hidden dangers in flood discharge channels downstream of the reservoirs. Pilot works should be initiated to clear sedimentation from silt-clouded reservoirs in large and medium size to restore and maintain the functions of flood regulation and storage. Full consideration should be taken to issues of reservoir inundation, relocation and resettlement. Studies and researches should be made to tap the flood control potential of existing reservoirs.

6.1.1.3 Promoting the Construction of Flood Storage and Detention Basins

Through the construction of flood storage and detention basins, the goal can be achieved of "Floods Being capable to Be Diverted into, Be Retained and Detained in, and Be Drained out". It is necessary to develop important flood storage and detention basins in advance and then the general and reserved one step by step.

The construction of the flood storage and detention basins should be sped up in vicinity of Chenglingji capable of diverting 10 billion m^3 of excess flood. Focus will be put on the security construction of such flood storage and detention basins as Qianglianghu, Gongshuangcha and Datonghu in Tongting Lake Area, as well as the construction of embankments in flood storage and detention basins in east part of Honghu Lake Area, and sluices for diverting flood into and releasing out of it, and the safety construction. The development of important flood storage and detention basins should be initiated, including the Minzhu and Chengxi polders. Actions should be taken to accelerate the construction of structures and security facilities in the

Dujiatai flood storage and detention basin near Wuhan City. The pre-investment studies should be started works for the construction of structures and security facilities of Wuhu and Zhangduhu detention basins, and the implementation of them should be promoted. The construction of security facilities should be accelerated in the Kangshan detention basin near Hukou City and the Huayanghe detention basin is to be developed. More attention will be paid to the population management and industrial restructuring in detention basins. It should be well developed for the utilization of flood resources from rivers and lakes and the ecological conservation of wetlands in detention basins. Studies should be made to adjust the layout of flood storage and detention basins according to the changes in flood control situation on the middle and lower reaches of the Yangtze River and corresponding to the requirements of its socio-economic development.

6.1.1.4 Strengthening the River Channel Regulation

The stability in river regime and the capacity in flood storage and discharge along river channel can be maintained through river regulation. As the river–lake relation changes and the river channel evolves after the completion of key reservoirs in the upper Yangtze River, the bank collapse rehabilitation and river regime control should continue to be strengthened for the mainstream of the middle-lower Yangtze River, and the systematic rehabilitation should be made for it too. Comprehensive rehabilitation should be conducted for the four tributaries of the Dongting Lake to maintain the capacity of flood diversion from the Jingjiang main stretch to the Dongting Lake. River regulation should be put into practice for the floodway flowing into Dongting Lake and lower channels of Xiangjiang, Zishui, Yuanshui, and Lishui rivers, the 4 tributaries of Dongting Lake, as well as Ganjiang, Fuhe, Xinhe, Raohe and Xiushui rivers, the 5 tributaries of Poyang Lake. The lower channel regulation of major tributaries such as Hanjiang River and Jialing River shall be implemented to enhance the discharge capacity of the floodway by measures of dredging, obstacle removing and bottleneck extending. While ensuring the stability in river regime and the effective function in flood discharge, attention should be specially paid to ecological restoration and protection.

6.1.1.5 Strengthening the Construction of Drainage Works for Local Floods

We should strengthen the governance of key areas such as Dongting Lake, Jianghan Plain, Poyang Lake and polder areas along river streams, and well protect water bodies with functions of regulation and detention such as the rivers, lakes and depressions, etc., so as to maintain the capacity in storage of local floods. Measures such as development of drainage pump stations, rehabilitation of canal systems, reinforcement of embankments and culvert sluices shall be taken to improve the capacity in drainage of local floods. Following the concept of "Sponge City", developments should be strengthened in urban water system connection improvement, river and lake dredging, and underground drainage system according to local conditions. The river basin flood control and regional drainage should be well coordinated, with large pumping stations discharging to the Yangtze River integrated into the flood control system on the middle and lower reaches of the Yangtze River.

6.1.2 Building a Flood Defense System

Modern technologies are applied to rainstorm flood monitoring and forecasting, flood regulation, flood control decision-making and flood control project management. The flood control works will play a better part through an optimized joint operation among projects as reservoirs and flood storage and detention basins.

6.1.2.1 Improving the Capacity in Forecasting and Warning Rainstorm Flood

Efforts shall be made to improve the development of hydrometric and meteorological networks, optimize the layout of the comprehensive monitoring station networks in the River Basin, develop innovative testing methods, and achieve information sharing in different industries and sectors such as meteorology, oceanography, transportation and water conservancy in the River Basin. Profound studies shall be made on the law of changes in hydrological regime incurred due to completion of water conservancy and hydropower projects. Revisions shall be made to improve the forecasting alternatives for rainstorm floods in control sections and water projects on the mainstream and tributaries. We should make active response to the impact of global climate change on extreme weather events and further improve the level of monitoring, warning and forecasting of hazardous weather such as rainstorms to prolong the prevision time and improve the accuracy of forecasting.

6.1.2.2 Ameliorating the Flood Defense Alternative System

Preparation shall be initiated for flood defense and flood regulation alternatives for major tributaries like Yalong River, Minjiang River, Dongting Lake and Poyang Lake. Preparing flood defense and flood regulation alternatives for major rivers will form a flood regulation alternatives system at level of the River Basin. Programs shall be formulated in detail to protect the River Basin against above-standard and extraordinary floods and programs shall be prepared for operation of important and ordinary detention basins, for which drills are exercised when appropriate.

6.1.2.3 Strengthening the Joint Operation of Key Projects

The mode of joint operation for flood control shall be optimized for group reservoirs in the mainstream and tributaries with the Three Gorges and the Danjiangkou reservoirs in nuclear role, and the optimization of joint operation plans shall be carried out for key water projects in the River Basin on a rolling basis. In-depth studies and researches shall be conducted on the mode of flood control operation of cascade reservoirs on tributaries. Taking the flood forecasting into account, a joint operation will be gradually realized in the whole basin covering key reservoirs, storage and detention basins, and backbone diversion and drainage projects.

6.1.2.4 Heightening the Prevention and Control of Flash Flood Disasters

In accordance with the principle of combining prevention with control and giving priority to prevention, continuous efforts shall be made to strengthen the prevention against and control of flash flood disasters, with focus on improvement of the capacity in monitoring and warning. We should refine measures for mass observation and preparation and consolidate and improve non-engineering measures for prevention against flash flood disasters. Actions shall be taken to control key flash flood gullies. Risk management of flash flood disasters shall be further implemented to identify and remove potential risks in residential areas as well as industrial and mining enterprises in order to avoid risks and prevent disasters.

6.1.2.5 Better Safeguarding Flood Control Operation with Informatization

Improvement shall be made for the joint operation system of water projects in the Yangtze River Basin, with the coverage gradually expanded and the targets broadened under regulation. In-depth studies shall be conducted to gradually realize IT-based joint operations featuring artificial intelligence applications. With deep integration and inheritance of system resources, the facilities will be set up in multi-source data collection platform, big database and supporting platform. We will also explore the application of frontier information technology in flood control operation system to improve the capacity in providing information service for flood control and disaster mitigation.

6.1.3 Establishing a Flood Risk Management System

Giving the top priority to ensuring people's life safety, we will raise the public awareness in flood control and disaster mitigation and popularize relevant knowledge and skills in self-help and mutual rescue. We should establish a strict flood risk management system under the prerequisite of respecting the law of nature, coordinate effectively the relationship between man and nature, and give the way to floods, so as to support the sustainable socio-economic development.

6.1.3.1 Strengthening Risk Management

We need to carry out the work of designating the flood discharge and storage space for rivers and lakes, which should be coordinated with the national spatial planning, and initiate studies on the requirements of water space control. Efforts are required to continue improving the formulation and application of flood risk maps, prepare risk zoning maps for flood and waterlogging disasters, promote socialized management on flood and waterlogging disasters, strictly control construction projects in high-risk areas, guide the flood control work and socio-economic development in the River Basin. We also need to carry out study on flood and waterlogging insurance systems, specifically, exploring ways for use of social funds to compensate for losses by flood and waterlogging. It is necessary to conduct public education and skill training on flood control and disaster mitigation, so as to further raise the public awareness and ability in prevention against floods and waterlogging.

6.1.3.2 Improving Emergency Management

We will improve the emergency response mechanism for water related incidents such as floods and waterlogging, sudden river bank collapses, and serious quality and safety accidents of water conservancy projects; revise emergency response program for flood control and strengthen skill training and emergency drills; strengthen the capacity building and management of rescue teams.

6.1.3.3 Formulating Management Systems of Detention Basins, Floodplains and Polders

On the premise of meeting the overall requirements on the flood control in the Yangtze River, the relationships should be properly handled between flood discharge, flood detention and self-protection in the detention basins, floodplains and polders. Any encroachment on flood storage and detention basins should be prohibited, especially urbanization activities and industrial parks outside the safety zones in important and general detention basins. The management and protection measures shall be formulated for floodplains and polders in the middle and lower reaches of the Yangtze River.

6.1.3.4 Studying and Improving the Mechanism of the Payment for Flood Discharge and Storage Services

Based on summarizing the experience of compensation to operation of flood storage and detention basins in recent years, we need to further study the long-term payment mechanism for services of detention basins and make revisions when appropriate to the Interim Methods on Payment for Services of Flood Storage and Detention Basins. We also should formulate payment mechanisms for services of flood discharge and storage by floodplains and polders, and study the payment-related objects, scope, standards, fund sources and procedures on the basis of specific situation of each floodplain and polder.

6.2 Comprehensive Water Resources Utilization

6.2.1 Promoting the Construction of a Water-Saving Society

We will fully implement the policy of giving the priority to water conservation, strengthen the rigid constraints raised by the carrying capacity of water resources, and accelerate the shift from extensive to efficient use of water. We must strictly manage water use throughout the whole process, with water conservation policy exerted in the entire procedure and in all fields of socio-economic development.

6.2.1.1 Dual Control in Total Volume and Intensity

More importance shall be attached to the rigid constraints raised by water-saving indexes. Classification shall be made for areas according to their carrying capacity of water resources to implement differentiated control. The intensity of water resources development and utilization shall be strictly controlled to determine urban spatial layout and scale, as well as industry scale and structure by availability of water resources, and also rationally determine the economic layout, structure and scale. We will sternly implement the water permit system. The management of water use should be strictly exercised throughout the whole process. Assessments will be performed according to the strictest possible system for water resources management, in the form of the strengthened target-oriented responsibility system and supervision assessment for water conservation.

6.2.1.2 Agricultural Water Conservation and Discharge Reduction

With the help of comprehensive water price reform for agricultural activities, we are fueled up to promote water conservation and emission reduction in the River

Basin. In detail, we will speed up to clarify agricultural water use rights, formulate water-conservation standard quota system covering major crops, advance the amount control and quota management on agricultural water use, establish agricultural water-saving evaluation mechanism, and improve the scientific management system and technical support for irrigation water use. The water-saving methods for animal husbandry and fishery should be facilitated. In rural areas, we will strengthen the construction and renovation of rural domestic water supply facilities and supporting pipe networks, and start renovation of rural domestic facilities, as well as promotion of water-saving appliances.

6.2.1.3 Industrial Water Conservation and Emission Reduction

We will focus on management of productive water uses, and vigorously promote water-saving technologies and techniques such as efficient cooling and washing, cycling use of water, reuse of reclaimed water, and substitutions of high-water-consumption technologies. We will encourage industries with high water consumption to conserve water and increase efficiency by guiding them to concentrate in industrial parks with appropriate water resources conditions. It is strictly forbidden to develop projects with processes, technologies and equipment on the elimination list but to knock them out as scheduled. We will actively promote cascade and cycling use of water and encourage existing enterprises and industrial parks to execute water-saving centered green and quality transformation, upgrading and cycling renovation.

6.2.1.4 Urban Water Conservation and Loss Reduction

To push forward water conservation and renovation in urban areas, we will accelerate the formulation and implementation of programs for transforming water supply networks, to substantially reduce the leakage of pipelines. We will make more efforts to save water in public areas and popularize water-saving appliances. Strict control should be made on water use in high-consumption service industries by actively promoting technologies, equipment and processes for recycling use of water. We also need to improve the level of rain water resources utilization and reclaimed water reuse, for optimal use of quality water and cyclic use in proper sequence.

6.2.1.5 Water-Conservation Evaluation

Water conservation should be properly monitored from the very beginning. To achieve this goal, evaluation should be carried out in terms of water-saving at the stages of plan observation and project approval. If the water conservation evaluation fails, related documents should not pass the technical observation such as the project plan, project proposal, feasibility study report and water resources verification report.

6.2.2 Optimizing Water Resources Allocation

We should work to optimize the pattern of river-basin water resources allocation, promote the construction of key projects, improve the capability in water resources regulation and water supply security, strengthen the upgradation of capability in emergency response to droughts and construction of urban complementary water sources for emergency, promote the formulation of water volume allocation programs for and integrated regulation of major rivers, so as to provide strong support for crucial strategy of national water resources allocation.

6.2.2.1 Optimizing the Pattern of Water Resources Allocation

Due to insufficient capacity in regulating water resources in the upper reaches of the Yangtze River, it is imperative to integrate the storage, diversion, and pumping measures, accelerate the efforts at pre-investment stage for major projects including Water Diversion from Dadu River to Minjiang River, Water Allocation in the Western Chongqing City, Long March Canal Diversion, advance the progress of Water Diversion in Central Yunnan Province, and give priority to solving water-shortage problems in central Sichuan Basin, Central Yunnan Plateau and Central Guizhou Plateau. In the middle reaches of the Yangtze River, a number of key water source projects should be built (renovated or expanded) based on actual conditions to further improve the regional water supply guarantee. We should speed up the construction of some major projects such as water resources allocation in Northern Hubei Province, and focus on solving the water shortage in Southern and Central Hunan Province, Southern Jiangxi Province, Tangbai River and Humpy Ground in Northern Hubei Province. In the lower reaches of the Yangtze River, such projects should be sped up as Water Diversion from the Yangtze River to Huaihe River. The improvement should be accelerated to form a multi-source complementary water supply pattern, and the monitoring and integrated scheduling should be strengthened for diversion and drainage projects along rivers for a corresponding operation in order. A guarantee system should be established with the characteristics of integration of large, medium, small and micro projects and dynamic combination of different water source works such as water storage, diversion, pumping and transfer, so as to basically achieve balance between the supply and demand of water resources in normal and medium-dry years.

6.2.2.2 Facilitating Flow Allocation for Trans-provincial Rivers

We will fully implement the formulation of water volume allocation schemes for major trans-provincial rivers, and on this basis, all provinces should speed up the water allocation for rivers running through cities and/or counties by refining and decomposing the total water use indexes in the basin into cities and/or counties, so as to basically achieve full coverage of water use limit in the River Basin. We will accelerate implementation of the approved schemes for water allocation in river basins and strictly control the total amount of water use in each region. Efforts will be put into strengthening the capacity building of hydrological monitoring and forecasting, exercising overall coordination for water demand over the upstream and downstream, and of various sectors and regions, programming annual water use plan, strengthening management of water drawing from rivers and strictly observing the flow through main river sections, enhancing management on regulation of water from conservancy and hydropower projects, improving the consultation mechanism for river water regulation, and intensifying supervision, inspection and accountability, so as to realize rational water resources allocation in trans-provincial river basins.

6.2.2.3 Establishing and Improving the System of Emergency Water Source Works

Relying on the integration of large, medium, small, and micro works, in accordance with local features, we should give priorities to maintenance, renovation and complementary facility development for existing water source works, expand and improve the water supply capacity and functions of existing water conservancy projects, and strengthen construction of connection works with water systems connected, operation integrated and mutual complementation formed. We also need to rationally arrange emergency water sources, optimize and integrate various water supply projects, and take overall measures to execute allocation and scientific regulation, aiming at improving overall capacity in emergency water supply.

6.2.2.4 Establishing and Improving the Response Mechanism for Emergency Water Supply

We will foster development of monitoring, forecasting and warning systems for water quality and quantity, establish and improve emergency plans for water pollution outburst and severe accidents due to extreme weather such as drought, and perfect water resources management and project operation programs under different drought situations. By strengthening the overall allocation and scientific regulation of water resources in river basins and regions, especially the joint operation of water resources on the right bank and left bank, mainstream and tributaries, upstream and downstream of backbone water resources allocation projects, we can maximize the emergency water supply to ensure the most fundamental water use of urban and rural residents, key industries, grain bases and basic grain ration fields. In case of special dry years, we should first ensure the basic domestic water use of residents, guarantee as much as possible the most basic water needs of key urban sectors, organizations and enterprises, and try the best to ensure the most basic water use of crops during the sowing period or the critical growth period.

6.2.2.5 Providing More Strategic Support for Water Resources Allocation

Endeavors should be made to ensure the safe operation and water supply security of the South-to-North Water Transfer Project, constantly enhance the management level of project operation, perform better for water regulation, and speed up the project acceptance and the construction of complementary facilities. In terms of water transfer, we should facilitate the construction of emergency water supply project for the first phase of the East Route to extend northwards to provide water for Beijing, Tianjin and Hebei Province, accelerate the efforts of justification at pre-investment stage for the follow-up tasks of the Middle and East Routes projects, the Middle Route Project which diverts water from the Yangtze River to the Hanjiang River and the regulation and storage projects along the route, and continue the efforts of justification at pre-investment stage on the West Route Project.

6.2.3 Improving the Backbone Network for Comprehensive Water Resources Utilization

In accordance with the principle of giving priorities to water conservation, protection and comprehensive utilization, we will work hard to make up for the areas of weakness in water conservancy infrastructures, and build a sound backbone network for water supply, power generation and shipping in river basins, so as to provide strong support and guarantee for major national strategies such as socio-economic development, rural revitalization and Yangtze Economic Belt development.

6.2.3.1 Improving the Backbone Network for Urban Water Supply

We will fully tap the water supply potential in existing facilities, accelerate the construction of key water source projects, and effectively solve the water shortage in key areas and cities in the River Basin. More work needs to be done to develop backup emergency water sources and multi-source supply systems in cities above county level and implement connecting works for water plant pipelines to realize mutual complementation of water sources. We strive to formulate and improve emergency response programs for water distribution in extremely dry years, which substantially improves the security of urban water supply systems and the capacity in responding to accidents.

6.2.3.2 Promoting the Integrated Urban and Rural Water Supply

The rural drinking water security will be consolidated and improved by the construction (expansion) of various drinking water source projects with functions of storage, diversion and pumping as well as water drawing and purification facilities and extension of water supply pipelines. In regions where conditions permit, we will promote integrated urban and rural water supply to extend water supply from cities to villages and towns, with overall plan on urban and rural water supply pipelines for unified construction, water price, water quality, standards, and services, to gradually allow the rural residents to enjoy the same quality, prices, networks and services as the urban residents and achieve efficient and quality operation of the water supply systems.

6.2.3.3 Consolidating Rural Water Infrastructure Networks

We should consolidate and improve the security of rural drinking water and enhance development of irrigation and drainage infrastructures, advance the capacity building scientifically orderly for key water conservancy projects, small and medium-sized water source projects and for responding to drought emergencies, build rural water conservancy infrastructure system with combination of large, medium, small and micro works and free connection of backbones with field terminals, develop complementary facilities, water-saving renovation and modernization construction for large and medium-sized irrigation districts, intensify efforts to upgrade water-saving complementary facilities for key medium-sized irrigation districts, exercise standardized management for irrigation districts and pumping stations, and qualify small irrigation and drainage facilities to provide support and guarantee for the rural revitalization strategy. We should speed up the development of irrigation districts including the first phase of Xiangjiaba Irrigation District, the second phase of Wuyin Irrigation District, the second phase of Shengzhong Irrigation District, and the first phase of Pihe Water Supply in Dujiangyan Irrigation District in Sichuan Province, the second phase of Liaofang Irrigation District in Jiangxi Province, the irrigation district of the expansion project of Centianhe Reservoir in Hunan Province and the Gangkouwan Irrigation District in Anhui Province. We will actively push forward the commencement of the Tingzikou Irrigation District and the second phase of Pihe Water Supply Project in Dujiangyan Irrigation District and speed up efforts for projects at pre-investment stage such as the complementary facilities in irrigation districts of Central Yunnan Water Diversion and for the Sichuan Long March Canal Diversion.

6.2.3.4 Developing Hydropower Resources in a Rational and Orderly Manner

On the premise of ecological conservation and resettlement of inhabitants, we will develop hydropower resources in a rational and orderly manner. With overall planning and rational distribution, we will accelerate the construction of pumping storage power stations in areas where conditions permit. To follow the Opinions on Sorting-out and Rectification of Small Hydros in the Yangtze Economic Belt Jointly Released by the Ministry of Water Resources, the National Development and Reform Commission, the Ministry of Ecology and Environment and the National Energy Administration, we will vigorously promote the sorting-out and rectification of prominent eco-environmental problems arising from small hydros, make sustained efforts to promote the green transformation with rural small hydros, strictly control small hydros with commercial purposes, and clarify responsibility in safety production for rural hydros. Those outstanding eco-environmental problems existing in small hydro development will be corrected to protect and restore the river ecosystem.

6.2.3.5 Accelerating Navigation Development

We will bolster the construction of the golden waterway, further improve the navigability of the Yangtze River mainstream, and enhance the strategic role of the Yangtze River to support the prosperity of the Yangtze Economic Belt. In the upstream regions, the key task is to improve the waterway conditions, mainly by means of cascade channelization to well rehabilitate the navigation-obstructing shoals and to practice corresponding maintenances. We will accelerate the justifications on favorable measures such as the new waterway of the Three Gorges Project etc., to ease the bottleneck problem in shipping along the middle stream. Rehabilitation and dredging will be carried out for downstream waterways, and efforts are made to strengthen the national network of high-grade waterways in the Yangtze River Delta and the channel dredging of coastal ports. We will work to shape a new pattern in which Shanghai International Shipping Center will take the lead, Chongqing Shipping Center on the upper Yangtze River, Wuhan Shipping Center on the middle Yangtze River and Nanjing Regional Shipping and Logistics Center will assume main responsibility. The main ports on the trunk or tributaries of the Yangtze River will serve as support with coordinated development among those ports on upstream, midstream and downstream, so as to build an open, beautiful and high-grade waterway network which links the river and the sea and connects the mainstream and the tributaries. We will accelerate the cascade channelization and rehabilitation of tributary waterways along rivers of the Minjiang, Jialing, Wujiang, Hanjiang, Xiangjiang, Yuanjiang, Ganjiang and Xinjiang and actively promote the development of waterways such as the Qujiang River, Songxi-Lizi waterway, so as to further drive the development of their hinterlands. Working with the East Route Project of the South-to-North Water Transfer and the diversion from the Yangtze River to the Huaihe River, we will accelerate the trans-river shipping connection. Studies shall be strengthened on the planning of trans-river canal routes and shipping works of Jiangxi-Guangdong, Hunan-Guangxi, and Zhejiang-Jiangxi areas and development shall be launched when appropriate for trans-river connection works.

6.3 Eco-environmental Protection

6.3.1 Intensifying Water Resources Protection and Water Pollution Control

We will set up the bottom line management system of water environment quality based on the water function zoning system, adhere to the strategy of comprehensive prevention and control of point, non-point and endogenous source of pollution, highlight the protection of centralized drinking water sources and the control of serious water pollution, strengthen the control of total phosphorus pollution, solve the outstanding water environment problems, strictly protect against severe risks in water environment, and practically maintain and improve the water environment of the Yangtze River.

6.3.1.1 Securing Ecological Water Use

We will coordinate the relations between upstream and downstream and between mainstream and tributaries, perfectly ameliorate the operation management and optimized regulation of river–lake system connection, and effectively increase the discharge in dry seasons. All these measures help ensure ecological water use in mainstream and tributaries of the Yangtze River, maintain the water level of ecological needs in lakes, and facilitate the steady restoration of river and lake ecosystems.

6.3.1.2 Protecting Centralized Sources of Drinking Water

We will demarcate protection zones surrounding centralized drinking water sources, remove illegal buildings in Grade I water source protection zones, ban all cage farming activities in Grade II water source protection zones, carry out rehabilitation activities to effluent outlets and wharves that affect the safety of water drawing and strictly investigate and clear illegal activities in protection zones. Rehabilitation shall be carried out for centralized drinking water sources that do not satisfy the Grade III standard compliance. As for the water sources that are hard to meet the Grade III standard in short time due to upstream inflow or natural background influence, it is necessary to take measures to strengthen the treatment technologies in the water plants or change water sources to ensure drinking water safety.

6.3.1.3 Handling Heavily Polluted Water Bodies

Focusing on rehabilitation of black and smelly waters in urban areas and the hydrological unit where controlling sections are located with water quality worse than Grade V, we will fortify the effluent collection and treatment and comprehensive rehabilitation of the waters, and implement measures including source control and pollutant interception, water saving and effluent reduction, endogenous pollution rehabilitation, water replenishment to river channel, dredging for environmental purpose, restoration for ecological function, and sewage treatment for artificial wetlands, in a bid to facilitate comprehensive rehabilitation of rivers with water quality worse than Grade V, such as Fuhe, Fuxi, Jingshan, Nanfei, Paihe, Tanglangchuan, as well as pollutant interception works for such tributaries as Minjiang, Tuojiang, Jialing, and Qujiang. In accordance with the Master Plan of Layout of Water Intakes, Effluent Outlets, and Emergency Water Sources along the Yangtze River in the Yangtze Economic Belt, we will take engineering measures to rehabilitate the effluent outlets of cities above the prefecture level, with such engineer measures as construction under standard-compliance, deep treatment to tail-water, and merger, closure or relocation of effluent outlets.

6.3.1.4 Keeping Total Phosphorus Pollution Sources Under Control

Total control will be implemented for areas where the total phosphorus exceeds the standard, mainly covering Buyi and Miao Autonomous Prefecture and Miao and Dong Autonomous Prefecture in Guizhou Province, Qianjiang, Jingmen, and Shiyan cities in Hubei Province, Changde, Yiyang, and Yueyang cities in Hunan Province, Nanchang and Jiujiang cities in Jiangxi Province, Yibin, Luzhou, Meisan, Leshan, Chengdu, Deyang, and Ziyang cities in Sichuan Province, and Yuxi City in Yunnan Province. We will promote on a centralized basis the prevention and control of total phosphorus pollution in the river sections including the Three Gorges Reservoir Area and its upper reaches, the mainstream sections of the Yangtze River in Hunan and Hubei provinces, and the Tuojiang, Minjiang and Wujiang rivers.

6.3.1.5 Restoring Ecology for Key Lakes in Eutrophication

We will strengthen the restoration and rehabilitation of lakes with serious eutrophication in the River Basin. In the Chaohu Lake basin, focus will be placed on pollution control and ecological restoration, with increasement of ecological land use in the clean water runoff area in the southwest region, and prevention and control of agricultural non-point pollution can be by the construction of ecological ditches/canals and the reduced use of pesticides and fertilizers. In the eastern region, the lakeside ecological buffer zone will be built to maintain the water delivery passage. Comprehensive rehabilitation and ecological restoration will be implemented on major inflowing rivers to Chaohu Lake, such as Nanfei, Paihe, Tangxi, Shuangqiao and Shiwuli rivers to reduce pollution load into the Lake. We will continue promoting ecological restoration in Dianchi Lake basin, with more efforts on water diversion and nitrogen and phosphorus control. Priorities should be given to the source control and pollution interception as well as the rehabilitation of inflowing rivers in the northern Lake basin. Motor fishing boats and cage farming will be forbidden in Dianchi Lake, and measures are implemented including returning reclaimed lands to forests and grasslands, returning ponds to lakes, and returning houses to wetlands, together with application of bio-bacterial fertilizers, organic fertilizers and optimized and balanced fertilization technology on nitrogen control and phosphorus reduction. Comprehensive rehabilitation and ecological restoration will be carried out for such major rivers as Haihe, Wulong, and Daqing rivers to reduce the pollution load into the lake.

6.3.1.6 Improving Rural and Agricultural Environment

With focus on the areas with national major water diversion projects located and the areas surrounding the centralized drinking water sources, we will work hard to comprehensively improve the rural environment, with emphasis on rural waste treatment. We will speed up the treatment to domestic sewage and improvement of latrines in rural areas by multi-means of urban pipe networks extension, centralized treatment and decentralized treatment. Construction will be carried out on ecological-friendly rivers, lakes, ponds and dams and comprehensive rehabilitation will be initiated for rural watercourses, with focus on connection among water systems, renovation of bank slopes, interception and control of pollution, dredging and cleaning-up of river channels. We will accelerate the development of ecological-friendly and clean small watersheds.

6.3.1.7 Strictly Protect Against Severe Water Environmental Risks

We will work hard to push forward the risk prevention and control for centralized drinking water sources and develop backup water sources projects, in a bid to strengthen protection of drinking water sources for the South-to-North Water Transfer Project. Centering on prevention, we are striving to build a system for environmental risk prevention and control with enterprises as the main body, optimize the industrial layout, strengthen coordination and linkage, enhance emergency rescue capabilities, implement the whole-process management and control, and effectively deal with severe water environment risks in key areas. We will put in place the whole-process supervision on toxic and hazardous substances to protect against the risk of secondary water environmental accidents induced by transportation.

6.3.2 Strengthening the Protection and Restoration of the Water Ecological Environment

Adhering to the principle of giving priority to protection and focusing on restoration by nature itself and following the law of nature, overall planning will be developed scientifically based on local conditions to strengthen protection and restoration for key areas and important habitats, speed up ecological corridor construction and biodiversity protection, improve comprehensively the ecosystem stability and ecological service function, build a multivariate and symbiosis river basin ecosystem, and protect against the invasion of exotic pests.

6.3.2.1 Deepening the Management and Protection of Key Ecological Function Zones

We will deepen the management and protection of key ecological function zones by strictly sticking to the ecological protection red line, environmental quality bottom line, and resources utilization upper limit, preparing and implementing a negative list for industrial accesses to key ecological function zones, and strengthening ecological supervision over development and construction activities. In line with the ecological protection red line in the Yangtze River Basin, we will optimize and adjust the main functions and spatial distribution of protected areas, construct cluster nature reserves and protective communities for important rivers and lakes, forests, grassland ecosystems and aquatic lives, natural relics, minimal-population wild plants and critically endangered wild animals, and exercise strict protection and management. A dynamic monitoring system of air-ground integration will be established for nature reserves in the watersheds to comprehensively improve the systematic, refined and informatization-based management of nature reserves. We will strengthen the overall strategies for planning, construction and management for various protected areas including scenic and historic sites, natural and cultural heritage sites, and forest parks.

6.3.2.2 Accelerating the Construction of Green Ecological Corridors

We will work to restore the degraded forests, farmlands, lakes and grasslands, and accelerate the implementation of projects for protecting natural forests, returning reclaimed lands to forests and grasslands, returning grazing lands to grasslands, returning farmlands to lakes and wetlands, protecting wetlands, and restoring desertification. Taking the construction of the Yangtze River protection forest as the main component, we will develop protection forest systems along rivers, along roads, around lakes, and surrounding cities, strengthen the development of green pathways and farmland-forest networks, build ecological corridors with reasonable allocation, stable structure, and sound functions along the mainstream and tributaries of the Yangtze River, and promote afforestation synergy among mountains, plains, rivers and lakes, cities, and villages. We will also build the protection forest system for Two Lakes and One Reservoir and improve the system development in key areas to reinforce the ecological function of forest corridors.

6.3.2.3 Reinforcing the Protection and Restoration of River and Lake Wetlands

We will strengthen the protection of river and lake wetlands, prohibit reclamation of lakes, reinforce protection of plateau wetland ecosystems, and increase the area and protection rate of natural wetlands. We will carry out ecological investigations and health assessments on rivers and lakes and build up the construction of ecological security systems for key lakes and reservoirs such as Dongting Lake, Poyang Lake and the Three Gorges Reservoir. Lake and wetland ecosystems will be restored through regulation of water volume, lakeside ecological restoration, ecological water replenishment, river–lake connection, and restoration of important habitats. Comprehensive management of wetlands will be carried out through riverbank restoration, wetland vegetation restoration and pest control. It is encouraged to initiate pilot attempts of the payment for benefits from ecological services of wetlands, and practice of returning farmlands to wetlands. We will engage in wetland protection and restoration projects for international and national major wetlands, wetland natural reserves and national wetland parks, to gradually restore the ecological functions and expand the area of wetlands, so as to enhance wetland protection and management.

6.3.2.4 Intensifying the Protection of Biodiversity

We will strengthen the management of priority areas for biodiversity conservation in the upper reaches of the Yangtze River, and improve facilities for ex-situ protection of biodiversity. We will carry out surveys and assessments on ecosystems, species and genetic resources in key regions, and establish biodiversity databases and information platforms in river basins. Regular observation, monitoring, evaluation and warning for important biological groups and ecosystems, as well as national key protected species and their habitats will be carried out. We will strengthen the protection of biological and genetic resources and build conservation bases and gene banks for artificial populations of wildlife. We will facilitate the base construction for ex-situ protection of Yangtze River finless porpoise in the lower reaches of the Yangtze River to form ex-situ protection area groups. We will build rescue and breeding centers for rare and unique wild animals and perform pilot demonstrations to reintroduce artificial populations into the wild. Efforts will be made to perfect the management mechanism of breeding and releasing, establish and improve the tracing system of released germplasm, and strengthen tracking and evaluation on releasing effect.

6.3.2.5 Promoting the Ecological Restoration Project for Important Habitats

We will strengthen the restoration of migration routes for rare, endangered and unique fish resources in the mainstream and tributaries of the Yangtze River, build a decision-making supporting system for fish passage facilities in the Yangtze River

Basin, and implement the restoration projects for migration routes in the mainstream and major tributaries of the Yangtze River. More work will be done to strengthen important habitat restoration for spawning grounds, feeding grounds and wintering grounds (so called 'three grounds'), and carry out technical researches and demonstrations on river re-naturalization. We can restore the in-river and bankside habitats by connection of river–lake systems, restoration of river morphology, building deep pools and shoals, surface hardening removal and construction of artificial fish nests, continue promoting the connection of river–lake water systems on the middle and lower reaches of the Yangtze River, and restore river natural habitats and habitats of important aquatic organisms. We will carry out overall and comprehensive regulation based on the needs of aquatic organisms and taking into account other important functions, restore the hydrological conditions of habitats where important species such as Four Farm Fishes live and reproduce, and establish mechanisms to ensure the ecological flow at important cross-sections of rivers and lakes in mainstream and tributaries. Rescue protection will be practiced for the habitats of rare and endangered animals and plants, such as pandas, golden monkeys, finless porpoises, and orchids and the habitats of tiny populations of wild animals and plants.

6.3.2.6 Improving the Protection Capacity for Water Ecology Security

We will improve the capacity to ensure water ecological security and protect against major water ecological risks. A system featuring intelligent monitoring, warning, and prevention and control management for water ecological risks, will be set up based on micro-fluidic chips, Internet of Things and in-situ microscopic image recognition. We need to formulate methods for prevention and control of invasive alien species in the Yangtze River Basin, improve mechanism in bio-safety customs inspection, intensify inspection and quarantine of imported goods and conveyances, prevent the introduction of alien invasive species along with goods, conveyances and ballast water, and regularly publish the distribution of alien invasive species. The prevention, control and management of key invasive species shall be fortified. We shall make more efforts to study and develop adaptive prevention and control technologies for the transmission of nutritive salt, organic pollutants and heavy metals along the food chains and their biological accumulation risks. We will advance the construction of ecological security shelter in Sichuan Province and Yunnan Province, and optimize the ecological security shelter system so as to improve the quality and stability of ecosystems.

6.3.3 Strengthening the Prevention and Control on Water and Soil Erosion

We will earnestly implement the decision and deployment which are put forward by the Party Central Committee and the State Council, strictly enforce the Law of Water and Soil Conservation, stay firmly with the underlying principle of water conservancy reform and development characterized by making up for short boards in water projects and strengthening supervision in water sector. We will always adhere to the general requirements on soil and water conservation work featuring more efforts on measures in supervision and addressing weak links in rehabilitation. Centering on more supervision over water and soil loss induced by human activities, we need to speed up the rehabilitation of key areas in the River Basin, make solid progress in soil and water conservation monitoring work, and fully exercise relevant responsibilities for solid foundation, powerful supervision and effective governance, providing strong support for accelerating the construction of ecological civilization and for ensuring sustainable socio-economic development.

6.3.3.1 Intensifying the Supervision Over Water and Soil Erosion Induced by Human Activities

We are required to protect the ecological environment with the strictest institutions and the most rigorous legal system, formulate and implement the methods for assessment of performance in target responsibilities of local governments in water and soil conservation, and specify the main responsibilities of local governments in this regard. Slopes above 25° should be prohibited to be reclaimed for the purpose of crop cultivation. For economic forests planted on slopes above 25°, we should select the trees in a scientific way, rationally determine the planting scale, and take measures to prevent water and soil erosion. We will encourage the governments at the county level to define and announce the limits of land reclamation on slopes, urge them to return reclaimed slope lands to forests and grasslands and take measures for water and soil conservation. We will also strictly limit and ban approvals for production and construction projects within the range of ecological protection red line and accelerate the natural restoration of forest and grass vegetation in hilly areas. A multi-sector linkage mechanism led by the water and soil conservation department should be established to form a strong joint force for prevention and control of artificial water and soil erosion.

6.3.3.2 Accelerating the Comprehensive Rehabilitation of Water and Soil Erosion in Key Areas

To fortify the comprehensive rehabilitation of water and soil erosion in the upper reaches of the Yangtze River, we should stick to the crucial strategies of promoting well-coordinated environmental conservation and winning the final phase of the war on poverty alleviation, and accelerate the rehabilitation of water and soil erosion in key areas including the lower reaches of Jinshajiang River, the upper reaches of Jialing River, the Three Gorges Reservoir Area, the Danjiangkou Reservoir Area, and the stony desertification region in Southwest China. At the same time, the management on slope cultivated lands should be intensified. Priorities should be given to the water and soil erosion control projects of slope cultivated lands for ecological protection and rehabilitation. It is imperative to coordinate relevant departments to focus on the areas with concentrated slope cultivated lands and prepare overall plan for quality and high-standard control and intensive utilization of soil and water resources. Guided by the strategies of zonal prevention and control based on local conditions, we should establish and improve the water security systems such as water sources for agricultural irrigation, irrigation water conveyance and distribution as well as drainage for existing sloping land of 5–15° suitable for cultivation, build high-standard farmlands, and improve the quality level of cultivated lands and the comprehensive land productivity. For the existing sloping cultivated land of 15–25°, soil and water conservation measures should be taken according to local conditions and the appurtenant water systems and roads are improved, with plantation of special economic forests and fruits and in the way of intensive management. For the sloping cultivated lands over 25°, it is imperative to return the cultivated lands to forests and grasslands, promote ecological and natural restoration, and speed up the pace of sloping land rehabilitation. The comprehensive management of watersheds should be further strengthened. Guided by the objective of building beautiful and livable villages, we need to center on the strategy of rural revitalization to implement on an all-round basis the watershed management in a broader dimension and make overall plans to improve the management of mountains, rivers, forests, farmlands, lakes and grasses.

6.3.3.3 Promoting the Monitoring of Water and Soil Conservation in the River Basin

We should continue conducting the dynamic monitoring in key soil erosion control areas at national and provincial levels, fully understand the dynamic changes of soil erosion and the situation of water and soil conservation, and regularly issue announcements on water and soil conservation to the public. An information system covering all production and construction projects should be established for dynamic supervision over the whole process of the projects, so as to accurately identify problems in violation of laws and regulations, and to provide basic information support for the supervision of soil and water conservation.

6.4 Integrated River Basin Management

6.4.1 Improving Laws and Regulations

Laws, regulations and related policies are the basis of the coordinated protection of the Yangtze River and the foundation of river basin management. We are obligated to accelerate the legal system development for the Yangtze River Basin, such as Yangtze River Protection Law, by incorporating coordinated protection for the Yangtze River and the ecological civilization construction proposed by General Secretary Xi Jinping into law to comprehensively guide the rehabilitation and protection of the Yangtze River Basin. We will actively promote legislation on joint operation of key water projects in the basin, water allocation in the Hanjiang River basin, riparian protection of the Yangtze River, protection for water resources of the Yangtze River, management of floodplains in the mainstream watercourse on the middle-lower Yangtze River, and the Danjiangkou Reservoir management, so as to regulate and restrict all kinds of water-related activities. We will establish laws and regulations for flood management, improve relevant institutions for mediation and handling of water-related disputes, study and perfect the management systems for safeguarding the ecological water demand, demarcating the spatial range of rivers and lakes and the management scope of water projects, and building a basin-wide water-saving society.

6.4.2 Innovating Institutions and Mechanisms

Innovation shall be encouraged in the rehabilitation and protection mechanisms of the Yangtze River to execute the sector-decentralized management under the overall coordination mechanism at the level of the Yangtze River Basin. A coordination agency for protection of the Yangtze River shall be set up under the State Council, consisting of a technical advisory committee and an executive office. Relevant departments of the State Council should, in accordance with the division of responsibilities, strengthen the administration and management, exercise law enforcement, and assume corresponding responsibilities in the Yangtze River Basin. Local governments at various levels and their departments concerned should, according to division of responsibilities, strengthen relevant administration and management, exercise law enforcement and assume corresponding responsibilities within their respective administrative regions.

It is necessary to establish a high-level coordination mechanism for river-basin management. We will explore the establishment of a discussion and coordination mechanism consisting of the National Development and Reform Commission and relevant ministries of the State Council, involving water resources, ecology and environment, natural resources, transport, and agriculture and rural affairs, to address problems as result of insufficient organization, coordination and supervision at the level of central government. Efforts shall be made to organize and draw up important strategies at level of the Yangtze River Basin, study and address major issues in the Yangtze River Rehabilitation and Protection, coordinate integrally relationships between various types of protection and development among regions and sectors in the River Basin, and strengthen the coordination, negotiation and collaboration among the administrative responsibilities of relevant departments subordinated to the State Council and local governments at all levels. We still need to organize the formulation of standards for the Yangtze River Basin protection, the development of monitoring network platforms, the information collection, processing and publishing, and emergency response, as well as strengthen the coordination and joint action in law enforcement among water-related authorities.

There is a need to establish a working mechanism for the rehabilitation and protection of the Yangtze River Basin. We will further strengthen the integrated river basin management, establish a mechanism of River (Lake) Chief System with clear responsibilities, orderly coordination, strict supervision and strong protection, and implement the responsibilities of local governments and enterprises, so as to form a joint effort to exercise coordinated protection. We need to further clarify the administrative responsibilities among the River Basin and the administrative regions as well as different sectors and establish a working mechanism for the rehabilitation and protection of the Yangtze River Basin featuring distinct authorization, clear division of responsibilities, norm-based acts, and coordinated operation. Continued efforts will be made to improve indicator system for supervision and management and standards for evaluation and control of water-related affairs and set up and perfect supervision system and list of rights and responsibilities to strictly limit the projects affecting the ecological security of the Yangtze River Basin. All these measures ensure that the socio-economic development keeps in line with the requirements for ecosystem protection in the Yangtze River Basin. The government and various enterprises shall assume main responsibilities in protection of the ecosystem in the Yangtze River Basin, and establish relevant incentives, indemnity and accountability mechanisms.

6.4.3 Strengthening the Supervision and Management

Unified monitoring. Taking watershed as a unit and relying on existing hydrometric stations, the monitoring stations of various departments will be coordinated integrally to plan water-related monitoring networks in the Yangtze River Basin. To be specific, efforts will be made to overall plan the existing water-related monitoring station networks set up by relevant ministries and commissions, local governments and organizations in the Yangtze River Basin based on the existing hydrometric stations, following the principle of non-repeated development and combined with the monitoring items and historical operations, so as to rationally lay out the integrated network and various professional stations in the basin. A unified water-related information sharing platform shall be set up and an inter-sector information sharing mechanism will be established in the Yangtze River Basin. On the basis of the unified monitoring network planning, a unified information sharing platform shall be built to collect and integrate all aspects of water-related monitoring information, indicators, techniques and evaluation methods, providing strong support for the decision-making of the Yangtze River protection. We will unify the technical standards for monitoring of water resources, water environment and water ecology, strengthen the summary and analysis of monitoring information, carry out assessment and warning based on monitoring information, and realize the unified publication of water-related information.

Integrated planning. To achieve the target of strengthening the integrated planning for water-related affairs in the Yangtze River Basin, we ought to take into account such factors as carrying capacity of resources, population distribution, economic layout, land use, flood and drought disasters prevention, and eco-environmental protection. What is more, we will make scientific layout for spaces of production, living and ecology, and rationally delimit the space for management and control on rivers and lakes. We should strengthen the guidance and rigid constraints role of plans and establish the evaluation and supervision mechanism for follow-up implementation.

Joint operation. Overall consideration shall be given to the needs in flood control, ecology, water supply (irrigation), power generation, navigation and emergency response to implement the joint operation of key water projects in the basin. With clear objectives, operation items, technical standards, conditions of joint operation as well as responsibilities and rights of all stakeholders. With information-sharing strengthened, integrated operation systems shall be created for the key water projects in the basin, to improve the capability in analysis and decision-making. Integrated operation schemes shall be formulated in a scientific way and attempts shall be made to establish payment mechanism for joint operation services from key water projects in the basin.

Unified supervision. More efforts shall be made on the development of index system for water-related supervision and management, and the criteria for evaluation and control, and on the supervision system and the list of rights and responsibilities. We will promote top-level design for informatization and improve the IT-based standard system. Actions will be taken to accelerate the integration and sharing of information resources, promote deep integration of informatization into businesses, and further facilitate the application of One Diagram to the businesses of integrated river basin management. It is imperative to establish supervision databases and information systems and improve the water administration law enforcement platform. By doing so, we can better improve regulative tools and capabilities by means of informatization. We will speed up the construction of a water monitoring system covering the whole Basin and promote full-closure water-related supervision consisting of monitoring, assessing, warning, disposal and summary. In accordance with the principle of combined full-time and part-time duties, with focus on full-time personnel, we will build full-time teams for comprehensive law enforcement, who will strengthen supervision over water-related matters during and after the event. A system is expected to be established for linking law enforcement in water administration with criminal justice, by means of which communication and coordination with public security, prosecuting authorities and judicial organs can be enhanced. We will establish a long-term mechanism for prevention and mediation of inter-provincial water-related disputes.

6.4.4 *Promoting the Social Co-governance*

We will establish and improve mechanisms for public participation to fully solicit opinions from local authorities, departments and the public, and mobilize the initiative and enthusiasm of the society to participate in the coordinated protection of the Yangtze River.

We will explore to establish payment mechanism for ecological services with fundamental studies and researches on the scope, standards system and financing of the payment for water ecological services in flood storage and detention basins, centralized drinking water sources, and key prevention and control areas with water and soil erosion. In addition, we need to well balance the benefit relationships between upstream and downstream, left bank and right bank, and among different watersheds and different industrial sectors, and coordinate the relationship between national fiscal transfer payment and local complementary payment.

We will give full play to the leading role of the government and energize the market to establish a multi-investment mechanism for both rehabilitation and protection. Greater efforts need to be made to promote inter-regional and inter-user water right trade, gradually popularize the public purchase of water-related services including water pollution rehabilitation and ecological restoration, and give full play to the role of water resource taxes in regulating water resources allocation. We will integrate and apply multi-sectoral funds and technologies and rationally allocate responsibilities to form a synergy of governance and protection.

6.4.5 *Adding Scientific and Technical Support*

We will foster the development of scientific research platforms and make good use of existing ones, with more efforts on major issues of watershed rehabilitation and protection, to popularize and transform scientific and technological achievements into practice. We will enhance the integration of scientific research strengths and actively plan the application submission of national-level platforms including National Key Laboratory of the Yangtze River Protection and Flood Control and Disaster Mitigation, National Key Laboratory of the Smart Yangtze River, National Water Security Technology Innovation Center of the Yangtze River Basin and the National Field Scientific Observation and Research Station, and the provincial-level platforms including Technology Research Center for Joint Operation of Group Reservoirs, Technology Research Center for Watershed Water Ecological Conservation and Restoration, Laboratory of Warning Simulation for Water Resources Risks. We will give full play to the role of the YRPSTIA, which will help make significant breakthroughs on major issues on the Yangtze River rehabilitation, development and protection. We will strengthen the building of think tanks for the protection of the Yangtze River, make innovations in capacity building mechanisms and promote the optimization and adjustment of the talents structure to foster a qualified and professional talents team.

Part II
Major Issues Under Spotlight

Chapter 7
Master Plan of the Yangtze River Basin

The Master Plan of the Yangtze River Basin is the plan for overall arrangement of development, utilization, conservation, protection and administration of water resources and prevention and control against water-related disasters in the Yangtze River Basin, and the blueprint for rehabilitation and protection of the Yangtze River. The CPC Central Committee and the State Council attached great importance to rehabilitation, development and protection of the Yangtze River. The Changjiang Water Resources Commission (CWRC), together with relevant state authorities and local authorities in the Yangtze River Basin, have been insisting on taking planning as the forerunner, water-related management as the means, engineering development as the foundation, and scientific and technical innovation as the support, to carry out omni-directional rehabilitation, development and protection of the Yangtze River, resulting in outstanding achievements.

7.1 Preparations and Revisions of the Master Plan of the Yangtze River Basin

The CPC Central Committee and the State Council paid high attention to the efforts in planning for the Yangtze River Basin. Since it was founded, CWRC has undertaken the main task of flood control and actively made various preparations for the basin planning. In accordance with the state arrangement, CWRC, in conjunction with relevant ministries and commissions under the State Council, carried out three rounds of preparations of and revisions to the Master Plan of the Yangtze River Basin abreast of the times by completing Key Points of plan for Comprehensive Utilization of the Yangtze River Basin (1959), Executive Report on Plan for Comprehensive Utilization of the Yangtze River Basin (1990) and Master Plan of the Yangtze River Basin

© Changjiang Press (Wuhan) Co., Ltd. 2022

Changjiang Water Resources Commission of Ministry of Water Resources of the People's Republic of China, *Report on Yangtze River Rehabilitation and Protection 2019*,
https://doi.org/10.1007/978-981-16-4927-1_7

(2012–2030) respectively at the beginning of the founding of the People's Republic of China, after the reform and opening up and at the beginning of the twenty-first century, providing a reliable foundation and essential basis for water development and administration in the Yangtze River Basin.

7.1.1 Report on Key Points of Plan for Comprehensive Utilization of the Yangtze River Basin (1959)

In 1952, the Chinese Central Government proposed that the general direction of water development should be "planning shift from local level to basin-wide level, works from temporaries to permanents, and efforts from negative damage mitigation to positive benefit promotion". After the severe flood in 1954, the Central Government decided to speed up the regulation and development of the Yangtze River and prepare the Yangtze River Basin planning as soon as possible. In 1955, the overall efforts were made on the Master Plan of the Yangtze River Basin. CWRC was responsible for the preparation of plan at level of the basin, with joint participation of the Ministry of Fuel Industry, the Ministry of Transport, the Ministry of Railways, the Ministry of Geology, the Ministry of Fisheries, the Ministry of Electric Power, Chinese Academy of Sciences, and the Ministry of Culture, and with assistance from invited Soviet specialists. In March 1958, at the Chengdu Meeting hosted by the Political Bureau of the CPC Central Committee, the then Premier Zhou Enlai delivered a report on the Three Gorges Project and the planning of the Yangtze River Basin, which was discussed and approved by the meeting. On April 5 of the same year, the Political Bureau of the CPC Central Committee officially approved the document of CPC Central Committee's Opinions on the Three Gorges Project and the Planning for Yangtze River Basin. The document formulated the principle of "integrated planning, comprehensive development, appropriate division of task, and implementation in stages" for the Yangtze River Basin planning efforts, and stated that "It must be combined with each other the seven kinds of relations between the vision and the near future, the main stream and the tributaries, among the upper, middle and lower reaches, the large, medium and small sizes, among the efforts in the flood control, power generation, navigation and irrigation, the hydro-power and the thermal power, and the power generation and utilization, and arranged specifically according to priorities and the sequence order based on actual situations".

According to the opinions of the Chinese Central Government, CWRC formulated the Key Points of Plan for Comprehensive Utilization of the Yangtze River Basin (1959). The 1959 version report, on the basis of a large number of investigations and in-depth analysis on the characteristics of the Yangtze River and for the goal of the greater, faster, better and more economical socialist development, put forward for the first time a systematic plan and a development strategy for the Yangtze River management, and sketched a grand blueprint for the Yangtze River Basin planning with the Three Gorges Project as the main function component. The plan determined that the tasks of the Yangtze River Basin were to eradicate flood disasters to protect the safety of the people and the industrial and agricultural production, to eliminate the drought and waterlogging disasters to guarantee the rapid development of agricultural production, to prevent soil and water erosion for the economic development in the mountainous areas, to fully develop hydro-power to provide a large amount of cheap electricity so as to promote the rapid industrial development and the technological transformation of the national economy; to improve water transport conditions to increase shipping capacity and reduce transport costs to facilitate physical communication and trading; and to pay attention to the development of aquatic products and improvement of water sanitation. In addition, in the spirit of coordinating all the activities of the nation as in one chess game, comprehensive consideration must be given to and appropriate arrangement should be made for the issues of water transfer from south to north and canals linking adjacent watersheds.

In the Plan the overall development program of the Yangtze River Basin was divided into five components as the following: the program for hydro complex project development with function in flood control and power generation as main purposes, the program for water conservancy development with function in irrigation and water and soil conservation as main purposes, the program for comprehensive utilization in plain and lake areas with function in flood control and drainage as main purposes, the program for river channel regulation on the mainstream and for the development of south-north canals with function in shipping as main purposes, and the program for water diversion to adjacent watersheds.

The basic principles determined in the plan and the seven relationships to be well solved have laid the basic direction for the subsequent planning of the Yangtze River Basin. The Plan clarified that flood control was the primary task of Yangtze River management and the strategic position of the Three Gorges Project in planning efforts for the basin.

7.1.2 Executive Report on Plan for Comprehensive Utilization of the Yangtze River Basin (1990)

In 1972 and 1980, the Ministry of Water Resources and Power organized two symposiums on flood control on the middle and lower reaches of the Yangtze River, which further clarified the standards and overall arrangements for flood control on the middle and lower reaches of the Yangtze River. After the third plenary session of the 11th Central Committee of the CPC in 1978, China entered a new period of development and the focus of national work shifted to economic development. Such development situation put forward new requirements to regulation and development of the Yangtze River.

In order to meet the demands under reform and opening-up policy and to hit the strategic target on quadrupling the total output value of industry and agriculture by the end of the twentieth century, the State Planning Commission, in December 1983, after the approval by the State Council, made arrangement to revise the Key Points of Plan for Comprehensive Utilization the Yangtze River Basin (1959). The State Planning Commission together with the Ministry of Water Resources and Power should be responsible for the revision, with joint participation of relevant ministries and administration such as the Ministry of Transport, the Ministry of Construction, the Ministry of Agriculture, the Ministry of Animal Husbandry and Fisheries, the Ministry of Geology and Mining, the Ministry of Forestry, the Environmental Protection Administration, as well as CWRC and the planning commissions and relevant departments in the provinces (municipalities and autonomous regions) within the Basin.

In December 1987, CWRC completed Key Points of Plan for Comprehensive Utilization of the Yangtze River Basin. In 1988, the Ministry of Water Resources together with the Ministry of Energy discussed the Plan and put forward the revised and supplemented Plan in December of the same year and submitted it to the Central Government. In June 1990, the national leading group on water resources and water and soil conservation convened a meeting for review on the Plan in Beijing, chaired by State Councilor Chen Junsheng. After review, the leading group agreed to the Report in principle. In September 1990, the State Council ratified in principle the revised and improved the Executive Report on Plan for Comprehensive Utilization of the Yangtze River Basin (1990).

On the basis of comprehensive investigation and analysis on the natural and socio-economic conditions in the Yangtze River Basin including topography, geomorphology, geology, hydrology, meteorology, land, forest, mineral, energy as well as agriculture, urban, industry, transportation, and the existing flood, waterlogging and drought disasters and soil erosion, water pollution, schistosomiasis disease, and so forth, in the Executive Report on Plan for Comprehensive Utilization of the Yangtze River Basin (1990), subject was analyzed on the trend of economic development based on the economic situation and resource conditions of the basin and the requirements were put forward for comprehensive development and utilization of water resources, the favorable influence of the Yangtze River water resources development and utilization was analyzed on the national economic development with three components involved: (1) plan for comprehensive utilization, including development and utilization of water resources, flood control, waterlogging control, hydropower generation, irrigation, navigation, water and soil conservation, river channel regulation on the main stream of the middle and lower reaches of the Yangtze River, south-to-north water transfer, aquaculture, city layout, water supply, water resources protection and environmental impact assessment, tourism and so on; (2) plan for main stream regulation and development with 3 sections divided: Jinsha River section, Yibin to Yichang section, middle and lower reaches of the main stream; (3) plan for regulation and development in major tributaries, 19 tributaries or water systems in total, including the major tributaries on both sides of Jinsha River, the Yalong River, the Minjiang River, the Tuojiang River, the Chishui River, the Jialingjiang River, the Wujiang River, the Qingjiang River, the Dongting Lake and its tributaries, the

Hanjiang River, the Poyang Lake and its tributaries, the Huayang River, the Wanhe River and Caizi Lake, the Qingyijiang River, the Shuiyangjiang River, the Chaohu Lake, the Chuhe River, the Taihu Lake, and the Huangpu River.

In this plan, on the basis of inheriting the principles in Plan (1959), subjects were added as the relationships to be solved between the demands and possibilities, the overall and the local. Urban water supply and water resources protection were added to the planning tasks, the subjects planned were more substantial and perfect, and embodied the spirit of "keeping pace with the times" in the content. The integrated plan consisted of 14 components of tasks, taking flood and waterlogging control in the middle and lower reaches of the Yangtze River Basin, hydro-power development, and navigation on the main stream as the three major tasks according to the requirements of socio-economic development at that time, with planning and study efforts emphatically made thereon. The important role of the Three Gorges Project was reiterated in the Yangtze River Basin rehabilitation and development, and the subjects were refined for middle and East Route of the South-to-North Water Transfer Project and the Yangtze-to-Huaihe Water Diversion Project. It was better coordinated the relationship of integrated plan with the development tasks and the planned alternatives and the reasonable arrangements were proposed for the main stem and tributaries. The water resources protection and environmental impact assessment were put forward with the Yangtze River protection brought into the plan.

7.1.3 Master Plan of the Yangtze River Basin (2012–2030)

Stepping into the twenty-first century, the CPC Central Committee put forward a series of strategic plans for building a moderately prosperous society in all respects, practicing the Scientific Outlook on Development, and constructing a resource-conserving and environmentally friendly society. The near-term targets drawn up in the Executive Report on Plan for Comprehensive Utilization of the Yangtze River Basin (1990) was basically hit and the key projects including the Three Gorges Project were fundamentally completed. In order to support sustainable socio-economic development, the State Council arranged a new round of revisions to the Master Plan of the Yangtze River Basin, to be carried out by CWRC and the relevant departments of the 19 provinces (autonomous regions and municipalities) within the Basin. The Master Plan of the Yangtze River Basin (2012–2030) was approved by the State Council in 2012.

In the Master Plan of the Yangtze River Basin (2012–2030), with the Scientific Outlook on Development as the guide and "maintaining health of the Yangtze River and promoting harmony between human and water" as the basic objectives, and in accordance with the principle of "promoting development under protection and implementing protection during development", The "Four Systems" were planned to strengthen flood control and disaster mitigation, comprehensive water resources utilization, water resources and water ecological environment protection, and integrated river basin administration. The river function zoning system was improved, the controlling indicators were proposed for key locations on the mainstream and tributaries, and the classification were adjusted for some of flood detention basins. Focused on strengthening protection, the mainstream of the Yangtze River was divided into three types of reaches in function, namely, the reach prohibited to activities of development of hydropower resources, the reach reserved for further planning and the reach for activities of development and utilization. The program for cascade development was rechecked and adjusted in the upstream mainstream, and the programs were also proposed for mainstream in terms of flood control, water resources and ecological protection, sand mining, shoreline utilization, and river management. Programs for 48 major tributaries and some lakes were planned.

A new concept of "maintaining health of the Yangtze River and promoting harmony between human and water" was proposed in this revision of plan. It was clarified that full play should be given to the service function on the basis of maintaining the ecological function of the Yangtze River. The Plan strengthened water resources and water environment protection, paid more attention to drinking water supply for both human and livestock in rural areas, to the rehabilitation of small and medium-sized rivers, and to the prevention and control of mountain torrent disasters. Programed in Four Systems, more clear functions were exhibited in various tasks, which served to facilitate the coordination among various tasks in the systems. It was clarified the concept of overall planning and comprehensive development of mainstream and tributaries and proposed the coordination of mainstream and tributaries in flood control, water resources allocation and water resources protection, under the overall program for the whole basin.

Years of practice in the management of the Yangtze River has proved that each version of the integrated plans for the Yangtze River Basin has captured the characteristics of the Yangtze River and adapted to the requirements on socio-economic development for different periods. The guiding mentality and principles have been advancing with the times, along with concept being updated, overall program being optimized, and planning system being improved over the process, which have effectively guided and promoted the rehabilitation, development, protection of the Yangtze River and supported the sustainable socio-economic development in the Basin.

7.2 Role of the Master Plan of the Yangtze River Basin

In the integral interests of the State and the whole Basin, the Master Plan of the Yangtze River Basin, through combination of disaster prevention and benefit promotion, comprehensive utilization of water resources, and coordination among the rehabilitation, development and protection to well coordinate the complex relationships between/among the long-term and near-term targets, demands and possibilities, the whole and the local, mainstream and tributaries, the upper, middle, lower reaches, the left and right banks, river and lake, and so on, is the top leader in the overall situation of rehabilitation, development and protection for the Yangtze River and the basic guide for the sectors and departments engaged in water-related activities within the Basin, playing an important role that other forms of plan could hardly supplant.

7.2.1 Master Plan Emphasizes Strategic Arrangements

Water is the source of life, the essentials of production and the basis of ecology. Water is the essential natural resource and strategic economic resource that concerns national economy and people's livelihood, and the base element of ecological environment. Development of water conservancy and elimination of water disasters have a bearing on human survival, economic development and social progress and have always been a major issue in running the country. With the transformation of society and the change of nature, the influence of human activities is gradually increasing on the natural water cycle. The water issue has evolved into a complex problem of nature, society and humanity and the water security has become the bottleneck of social development in different stages, which requires strategic arrangement. The utilization of water resources should not only satisfy the current demands, but also the requirements of long-term sustainable development. Engineering development needs not only to solve current problems, but also try to find solutions to long-term problems, to make strategic arrangements for implementation progress based on the integrated program of regulation and the requirements of socio-economic development. In this connotation, the Master Plan is irreplaceable by other relevant plans with planning horizon of 5–10 years.

The Master Plan is designed to program strategic arrangements to deal with the major issues of water security in basin-wide based on the trend and requirements of socio-economic development. The Master Plan of the Yangtze River Basin in 1959, in view of the superior indicators of the Three Gorges Project in the comprehensive utilization such as flood control, hydropower generation, irrigation, and navigation, and its significant influence on the whole river, clarified the dominant position of the Three Gorges Project in all efforts programed for the Yangtze River Basin, proposed the idea on south-to-north water transfer, programed the 70-cascade hydropower development on the mainstream and tributaries, and made arrangements for actively promoting the justification on the Three Gorges Project. In terms of flood control, the Master Plan, based on the development trend and the bearing capacity of national economy, proposed the overall arrangement for flood control in three stages, which served as guidance to the development and continuous improvement in the flood control system of the Yangtze River. The Executive Report on Plan for Comprehensive Utilization of the Yangtze River Basin in 1990, on the basis of the Key Points of Plan for Comprehensive Utilization of the Yangtze River Basin in 1959, further defined the key role of the Three Gorges Project in the flood control system, preliminarily identified the scope of the Middle Route of the South-to-North Water Transfer Project, proposed the implementation opinions for development on the mainstream. It was made clear that the Three Gorges Project was an important project to be developed in short term, with the short-term development priority to such projects as

Xiluodu, Xiangjiaba, Jinping, Pubugou, Pengshui, Goupitan, Zipingpu, Tingzikou, Gaobazhou, Shuibuya, Jiangya, Zaoshi, Xiajiang, etc., which served as a guidance to the orderly development of hydroelectric resource on the mainstream and tributaries and the flood control and river channel regulation after the 1998 flood event occurred in the Yangtze River. The Master Plan of the Yangtze River Basin in 2012 refined the tasks of the key group reservoirs on the mainstream and tributaries in the upper Yangtze in coordination with the flood control task of the Three Gorges Reservoir, determined the rate of water resources exploitation and utilization at major interested sections and the indicators of water use efficiency and the total quantity of water use in the provinces (autonomous regions and municipalities), serving as a fundamental base for conservation, utilization and protection of water resources in the Basin and demarcated the upper limit for the water resources utilization in the national economic development.

7.2.2 *Master Plan Emphasizes Sustainable Use of Water Resources*

Water is an important national resource. Rivers are important carriers of water resources with multiple functions. Different sectors and regions have different demands for development, utilization and protection of water. The efforts for protection, regulation and utilization of the Yangtze River involves various ministries and commissions including the National Development and Reform Commission, Ministry of Water Resources, Ministry of Transport, Ministry of Natural Resources, Ministry of Agriculture and Rural Affairs, Ministry of Housing and Urban–Rural Development, Ministry of Industry and Information Technology, Ministry of Ecology and Environment, and sectors including flood control, water supply, protection, power generation, navigation, fishery and so on. Both development and regulation activities and the operation of key projects involve the interests from and conflicts among purposes of flood control, water supply, power generation, navigation and protection, and protection activities involve protection of drinking water sources, effluent control from industrial and domestic sewage, non-point source pollution control, aquatic species protection, wetland protection, etc. The Master Plan, from a long-term point of view, involves analysis on the current demand and existing prominent issues in process of protection and development, overall consideration of subjects as reasonable development, efficient utilization, scientific administration and effective protection of water resources, formulation of reasonable planning targets, overall arrangement with scientific connotation and practical program for implementation. Various sectors and industries have to communicate and participate in the process of formulation of the Master Plan of the Yangtze River Basin to jointly comply with it and facilitate the realization of sustainable and efficient utilization of water resources.

Although the Yangtze River Basin is relatively abundant in water resources, the Master Plan still emphasizes conservation and protection of water resources. The total amount is limited for water resources development and utilization in the Basin, not only to guarantee the rights and interests of the current generation in water resources utilization, but also respect the rights and interests of the future generations. The ecological water use is also arranged to ensure the ecological water demand and achieve the harmonious development between man and nature.

According to the natural characteristics of the river and the requirements of the economy and society, the Master Plan applies overall arrangement for the demands from different sectors. During the preparation of the Plan, the relationship between regulation and protection has been properly handled through inter-sectoral coordination, and the dominant service functions and development tasks have been clarified for different sections of the mainstream and tributaries, for the purpose of the comprehensive utilization of water resources. According to the overall arrangement in the Master Plan, various comprehensive development and utilization tasks are reasonably distributed in key projects, with defined controlling indicators and coordinated requirements of different sectors for management and protection, resulting in optimal overall benefits of the Basin. On the upper reaches of the Yangtze River, for example, the main task is for power generation, however, the pressure from flood control in the middle and lower reaches needs to be released by arrangement of flood storage capacity in them and the tasks of navigation and water supply should also be taken into account. On the middle and lower reaches of the Yangtze River, the main tasks are flood control and maintenance of stability in river regime, however, the needs for navigation, shoreline utilization and water resources utilization should also be taken into account. In the Executive Report on Plan for Comprehensive Utilization of the Yangtze River Basin in 1990, the important role of the Three Gorges Project in the comprehensive management and development of the whole basin was defined. The Three Gorges Project has generated great comprehensive benefits after its completion under the guidance of the planning and possesses the potential to replenish water to the Middle Route of the South-to-North Water Transfer Project.

7.2.3 *Master Plan Emphasizes Systematic Regulation and Protection*

The Yangtze River is a complete, independent and unique geographical unit composed of the mainstream, numerous tributaries and lakes. Mountains, rivers, forests, farmlands, lakes and grasslands are a life community, while rivers are the backbones of a basin. Xi Jinping, General Secretary of Central Committee of CPC, has stressed that "we should seek for new management approaches from the systematic engineering and overall perspective… The building of ecological civilization must take all factors into consideration with good overall plans and multiple measures." at the Conference on Ecological and Environmental Protection. The

Master Plan, aiming at prominent challenges in river basin regulation and protection and analyzing the causes thereof, has systematically involved the programs for regulation and protection through a variety of measures with the concepts of addressing the symptoms and root causes, integrating land use and water use in planning, and synergizing both government and market driving forces. In terms of flood control, overall plans have been made for the construction of embankments, reservoirs, flood storage and detention basins and river channel regulation with more efforts for protection of flood discharge and storage space, flood risk management, and utilization of flood resources. As to the issue of water supply security in the river basin, by combination of structural and non-structural measures, emphasis are not only put on the conservation and optimized allocation of water resources and development of water source projects, but also on the development of water supply projects and the protection of water sources and the regulation and management of water projects. In terms of ecological environment protection, emphasis are not only put on the conservation of water resources to keep strictly the human activities including management and development within the carrying capacities of water resources, water environment and water ecology, but also on the protection priority to fully implement the ecological environment protection so as to make efforts to minimize the impacts on the ecological environment in various stages including engineering design, construction and operation and fully achieve the ecological and environmental benefits of water conservancy projects.

The inseparability of water resources in a river basin and the interdependence of economy, society and culture determine that the integrated river basin management should take the river basin as a unit to gradually realize the integration of river basin management and administrative region management. In general, the Master Plan of the Yangtze River Basin serves as the blueprint and guideline for regulation, development and protection of river basin, involves the strategic arrangements for conservation, utilization, protection and management of water resources and mitigation of flood and drought disasters, and the water security support for national land space plan and economic development plan within the basin.

7.2.4 Master Plan Emphasizes Inter-regional Coordination

A river basin forms a whole through the hydrological cycle. The water resources in the river basin are inseparable and closely related to the economic society and human activities. Therefore, the water management of the basin must take into account the overall coordination to well deal with the relationships between the basin and the regions, the upstream and the downstream, the left and right banks, and the mainstream and tributaries. In the Master Plan, the interrelation and the mutual influence of different regions are taken into account and integrated into the framework of the overall benefits of the river basin. The Master Plan, taking the river basin as a whole, makes verification and comprehensive comparison of various schemes, and determines the function and development and protection tasks of different river systems and sections.

According to the water resources distribution characteristics of the Yangtze River Basin and the adjacent river basins such as the Yellow River, the Haihe River and the Huaihe River, taking the Yangtze River as the strategic water source of the nation, and in consideration of the limitation of water resources carrying capacity in the water export regions, such strategic arrangements of water reallocation have been formulated in the Master Plan of the Yangtze River Basin including the South-to-North Water Transfer Project, the Yangtze-to-Huaihe Water Diversion Project, the Hanjiang-to-Weihe Water Diversion Project, the Yangtze-to-Taihu Water Diversion Project, and the Water Diversion to Central Yunnan, which, by engineering development, has supported the national important strategy and the urban and rural water supply security. This includes the coordinated development in the Beijing-Tianjin-Hebei region, development of the Yangtze River Economic Belt, and integrated development in the Yangtze River Delta region, resulting in a coordination of the Yangtze River Basin and the whole country. At the same time, the effects on water export area and water import area have been considered as a whole, and necessary compensation measures have been proposed.

In terms of flood control, the coordination between the upstream and the downstream, and the left and the right banks have been taken into account to achieve the goal of overall flood control and minimum damage of flood disaster in the Basin. According to the contributions and regional components of floods flowing into the middle and lower reaches of the Yangtze River, the flood control policy has been determined as of "overall consideration of flood discharge and storage with preference on discharge". For the flood in the upstream catchment area of 1 million km^2, reservoirs have been programmed in the canyons along the upstream reaches of the mainstream and tributaries to retain and store floods, the reservoirs' capacities for flood control were planned on tributaries to work with the Three Gorges Reservoir based on the contribution of each tributary, which have greatly mitigated the flood pressure from upstream on the middle and lower reaches, with fewer losses by reservoir inundation. Appropriate contribution to flood control is also made by the middle and lower reaches, that is, to fully use the Dongting Lake and Poyang Lake to regulate and store the flood, to heighten and reinforce the embankments to increase flood storage and discharge capacity in river channels, and to use the flood storage and detention basins to retain and detain the excessive flood. The space was delineated clearly for flood discharge and storage, with equal arrangement for the detention basins in the vicinity of Chenglingji reach of the mainstream in Hunan and Hubei provinces. In the Yangtze River governance, great importance has been attached to coordinating the relationship between rivers and lakes, and measures have been studied and formulated based on the principle of "win–win for both rivers and lakes".

In terms of water resources and water ecological environment protection, according to the requirements on protection, restrictive constraints such as function zonings for hydropower development, shoreline and sand excavation have been demarcated, and the overall arrangement have been optimized for water intakes, effluent outlets, port terminals and river crossing pass-ways, ensuring the safety of water source areas and protecting the water ecological space.

7.3 Prospect of River Basin Master Planning in the New Era

Since the 18th National Congress, the CPC Central Committee with comrade Xi Jinping at its core has paid great attention to the national water security and the development of Yangtze River Economic Belt. General Secretary Xi Jinping proposed consecutively the water management guideline of "prioritizing water conservation, balancing development with spatial distribution of water resources, practicing systematic governance and achieving government-market synergy" in the New Era and the development conception of the Yangtze Economic Belt characterized by "promoting well-coordinated environmental conservation and avoiding excessive development", which pointed out the direction for the Yangtze rehabilitation and protection. To ensure the protection of the Yangtze River, all sectors and local authorities need to make concerted efforts to systematically conceive the ideas and methods for comprehensive management of the Yangtze River, to plan on a holistic arrangement for the rehabilitation and protection of the Yangtze River, to well conduct the work on the revision of the Master Plan of the Yangtze River Basin. In accordance with the requirements which were put forward by General Secretary Xi Jinping at the Symposium on Deepening the Development of the Yangtze River Economic Belt, the five major relationships should be well handled: the long-term and the present, protection and development, the whole and the step-by-step, the construction and the management, and the overall and the local. The old thoughts and ideas should be changed for innovative ways and means, and the supervision and assessment should be strengthened on implementation of the planning.

7.3.1 More Emphasis on the Comprehensive and Systematic Governance of the Basin

Water is an important part of the natural ecological system and the human system of the river basin. The river basin realizes the hydrological cycle through the veins of rivers and lakes. The impact of human life on soil and water is eventually merged into the rivers and lakes, and the water ecological environment has the most direct influence on human. In view of the active role of water in the basin, water regulation should be regarded as the breakthrough point to drive the comprehensive river basin rehabilitation and protection.

Water ecological environment is an important part of the ecological environment, involving various factors and the whole river basin, therefore, studies should be carried out from the integrity of the water ecosystem and the systematization of the Yangtze River Basin to work out action plans for the water ecological environment protection and rehabilitation in the Basin and systematically arrange long-term work with enhanced relevance and coupling to gradually realize the shift from post-governance to preferential protection and build up the water ecological and environmental protection system with coordinated development of man and nature and the strictest possible system for water ecological environment protection.

We should respect, follow and protect the nature's course and ensure that mountains, rivers, forests, farmlands, lakes and grasslands are a life community, and coordinate the top-level design for protection and rehabilitation of the water ecological environment. We should establish and improve the strictest possible systems for water ecological environment, strengthen the rigid constraints on the water ecological environment, establish an early warning mechanism for monitoring, evaluation and assessment of the carrying capacity of the water ecological environment, limit the number of permits in areas where the water ecological environment is overloaded, and protect the water ecological environment with the strictest institution and the rigorous rule of law. With inter-department coordination and regional collaboration, we should comprehensively strengthen the water ecological environment restoration by means of defining the ecological space for strict use control, returning farmlands to forests and grasslands for conservation and protection of water sources, returning reclaimed lands to lakes and wetlands for increasing regulating and storage capacity, restoring natural river channels for improvement of river functions, rehabilitating water ecological environment for improvement of aquatic creature habitats, strengthening management of river and lake shorelines and sand excavation, and so on.

In view of the current prominent issues, it is necessary to strengthen the joint operation of key water-related works in the basin, effectively deal with the changes in the relationship between rivers and lakes, make elaborate efforts to improve the relationship between rivers and lakes, and rationally allocate water resources. On the premise of meeting the domestic water demand, priority should be given to the guarantee of basic ecological water for rivers and lakes with rational allocation of water for production use. The ecological protection should be strengthened in important ecological reserves and water source conservation areas and the water should be well protected in Danjiangkou Reservoir and Liangzi Lake with good water quality. We should also promote and strengthen the ecological restoration and rehabilitation of key rivers and lakes (Dongting Lake, Poyang Lake, Chaohu Lake, Dianchi Lake, etc.), plateau lakes and their flow-in channels. It is necessary to explore, promote and implement restoration of migration routes of aquatic lives in rivers, lakes and reservoirs, strengthen ecological protection of old river channels and floodplains, strengthen ecological rehabilitation and renovation of embankments and beaches, as well as river ecological protection zones and ecological isolation zones. In combination with the improvement of irrigation districts, we should strengthen the control of non-point agricultural sources, promote the development of beautiful rural areas, and gradually eliminate urban and rural black and smelly water bodies.

7.3.2 More Emphasis on the Support, Guarantee and Constraint of Water Resources

Following the National Action Plan on Water Conservation, in-depth efforts should be made to build a water-saving society and overall account should be taken to the shortage of water resources and the deterioration of the water ecological environment. Shifting from increasing supply to more focus on demand management and in accordance with the principle of "determining urban and rural spatial layout and scale, construction land use, population size, industry scale and structure based on water resources", we should strengthen the management over the water quota and planned water use, strictly determine the amount of water intake and use, improve water use efficiency, and ensure the rigid restraint role of water resources in socio-economic development. Vigorous efforts should be made to advance water conservation and discharge reduction in key fields such as agriculture and industry, adjust and optimize the industrial structure, and phase out backward production capacity. The strictest possible systems for water resource management should be implemented and the "Three Red Lines" be strictly observed to control both the total quantity and intensity of water use. We will strictly implement the responsibility and appraisal system for water resources management and strengthen coordination and consultation on basin-and-regional combined water resources management. The carrying capacity of water resources should be evaluated and a monitoring and early warning mechanism should be established. The minimum discharge of key monitoring sections and from major water projects shall be supervised. The allocation of water resources should be optimized in a scientific way. We will continue formulating water allocation schemes and water scheduling plans for major rivers in the Basin, improve distribution of water resources, and promote development of key water sources works and major water diversion projects, push forward supporting and water-conserving facilities in irrigated areas, strengthen integrated scheduling and management of water resources, and promote establishment of an inter-department and inter-regional coordination and collaboration mechanism for water allocation.

The comprehensive utilization and recycling of water resources should be further strengthened. Supporting facilities should be provided and improvement should be made to existing important water conservancy infrastructures according to the requirements of modernization to improve their quality and efficiency, expand the irrigation and water supply benefits of reservoirs, further develop the water supply potential of the first phase of the East Route and the Middle Route of the South-to-North Water Transfer Project, and increase the volume of northward water transfer.

We will deepen the optimized operation of existing projects and improve the comprehensive benefits of water resources utilization, water disaster prevention and control, and river and lake ecological protection.

7.3.3 *More Emphasis on Overall Planning and Coordination in the Basin*

The Master Plan of the Yangtze River Basin should adhere to the water governance guideline of "prioritizing water conservation, balancing development with spatial distribution of water resources, practicing systematic governance and achieving government-market synergy" in the New Era and follow the development idea of the Yangtze Economic Belt as "promoting well-coordinated environmental conservation and avoiding excessive development" throughout the whole process of the water management and integrate the idea into all aspects of the socio-economic development and the construction of ecological civilization in the Yangtze River Basin, well deal with the relationships between socio-economic development and water resources development and utilization, water development and eco-environmental protection, as well as the relationships between benefit promotion and damage mitigation, protection and development, upstream and downstream, left and right banks, mainstream and branches to give full play to the functions of the Yangtze River and comprehensive utilization benefits and continuously improve the safeguard ability of water security. The sustainable use of water resources will promote the sustainable socio-economic development of the Yangtze River Economic Belt and the whole country, thus, laying a solid foundation for maintaining a healthy Yangtze River and building a beautiful China so as to achieve the "Two Centenary Goals".

The plan of integrated management of river basin should dovetail with the national territorial space planning. On one hand, it should fully mesh and comply with relevant requirements of demarcation and management of the "Three Zones and Three Lines", deal with the relationship between development and protection, and optimize the planning scheme, to provide support of water resources and guarantee of flood control security for realizing the goal of planning and make strategic arrangement of water demand for national territorial space development and protection for a long period of time. On the other hand, from the perspective of basin overall planning and according to the requirements of conservation and sustainable utilization of water resources, it should formulate the evaluation on water resource carrying capacity and the upper limit of water resource utilization and strengthen the constraint of water resources on socio-economic development. The management scope of rivers and lakes should be delimited and the requirements for shoreline management should be defined to maintain the main ecological functions of rivers, lakes and reservoirs, such as flood discharge and storage, water resources conservation and ecological breeding. To guarantee public security in flood control and water supply, it should also propose the development space, to be reserved in priority for water conservancy infrastructure development in a long period of time in the future.

7.3.4 More Emphasis on Supervision and Management

The management of water-related affairs should be comprehensively strengthened with the principle of "equal emphasis on development and management". With comprehensively strengthening supervision and management of water-related affairs, a fundamental shift will be made from emphasized development with neglected management to equal emphasis on development and management. While accurately strengthening the weakness in water resources utilization, flood control and disaster mitigation, water eco-environmental protection, water and soil conservation, and informatization and promoting development of water conservancy projects, more attention should be given to the supervision and management of water-related affairs to improve integrated river basin management and achieve a unification of project development and water-related management. A joint operation should be strengthened for key water works in the basin, the scope of such projects should be expanded, and the benefits of existing projects should be promoted.

In order to give full play to the guiding role of the Master Plan, it is necessary to establish a supervision mechanism for the implementation of the Master Plan. The specialized planning shall be subject to the Master Plan of the river basin, and the examination, approval and implementation of a large number of specialized planning in the basin, such as energy, transportation, urban construction and agriculture shall be subject to the Master Plan of the basin. Planning appraisal is an important measure to ensure effective implementation of the planning. In the process of planning implementation, the objectives, organization and implementation procedures of the planning as well as the social, economic and environmental benefits generated shall be comprehensively and systematically assessed to summarize experience and lessons in time, promote the effective implementation, and address comments and suggestions for adjustment and revision of the planning. Due to the different priorities and tasks of the basin protection, regulation and development in different periods, the revision to the planning should highlight the characteristics of the time to earnestly implement the new concept of water management and national strategic layout and meet the new requirements of ecological civilization development and "promoting well-coordinated environmental conservation" in the New Era to coordinate the Master Plan of the Yangtze River Basin with the national planning system.

7.3.5 More Emphasis on Science and Technology Leadership

Studies should be strengthened on major strategic issues including flow-sediment change, flood control standard verification, change of river–lake relation and regulation, water conservation standard in river basin, joint operation of water works and potentials thereof, development of strategic water source reserves and demonstration of Yangtze-to-Hanjiang Water Diversion Project. Studies and application of efficient utilization of water resources should be fostered to provide technical support for planning. Modern scientific and technological means, such as the Internet of Things, big data and GIS, should be applied and the traditional way of planning with texts and tables should be abandoned. The technologies, including numerical simulation, data mining, should be applied to innovate planning preparation, achievement display and ways of public participation. Space planning information linkage platform should be used to link up with the space planning.

It is necessary to innovate the organization form of the planning preparation and open the process. The Master Plan of the Yangtze River Basin should be revised and compiled democratically, openly and scientifically in accordance with the principles of government organization, inter-sector cooperation and collaboration, expert consultation and public participation. We should give full play to the role of experts and listen to the voices from experts in related sectors. We should make full use of various social resources and basic researches, give full play to their different advantages to participate in the planning, and listen to and absorb their views. We should solicit and listen to the opinions of relevant departments, bodies and the public through various channels and in various forms to expand the public participation. By thus doing, the planning can gather the consensus of the whole society and become a guide for water-related activities, with authoritativeness, guidance and operability.

Chapter 8
Change of River–Lake Relation and the Regulation of Dongting and Poyang Lakes

The Dongting Lake and Poyang Lake (hereinafter referred to as the Two Lakes), lie on the south bank of the middle reach of the Yangtze River and are the two largest freshwater lakes connected to the Yangtze River. The Dongting Lake has an area of 2625 km^2 when the water level is 34.4 m at Chenglingji station, with a storage capacity of 16.7 billion m^3. The Poyang Lake has an area of 3706 km^2 when the water level is 22.5 m at Hukou station, with a storage capacity of 30.2 billion m^3.The profiles of the Yangtze River and the Two Lakes and their tributaries are shown in Fig. 8.1.

8.1 River–Lake Relation

8.1.1 Relationship Between the Yangtze River and the Two Lakes

The Dongting Lake receives water from three outlets connected to the Yangtze River, namely Songzi, Taiping, and Ouchi outlet. There were once four outlets, however the fourth one named Tiaoxuan was blocked and gated in 1958. It is also fed by four tributaries, namely Xiangjiang, Zijiang, Yuanjiang and Lijiang River, merged with the Yangtze River at Chenlingji, forming a landscape of swallowing and spitting. The Poyang Lake is fed by five tributaries, namely Gangjiang, Fujiang, Xinjiang, Raojiang and Xiujiang River, and flows to the Yangtze River northward at Hukou. Back water effect usually occurs between the Poyang Lake and the Yangtze River, and waters from the Yangtze River sometimes flow backward to the Poyang Lake. The water regime variation of the Yangtze River directly affects the water volume of the Poyang Lake. The characteristics of water and sediments exchange between the Yangtze River and the Dongting Lake and the Poyang Lake make distinctive and

© Changjiang Press (Wuhan) Co., Ltd. 2022

Changjiang Water Resources Commission of Ministry of Water Resources of the People's Republic of China, *Report on Yangtze River Rehabilitation and Protection 2019*,
https://doi.org/10.1007/978-981-16-4927-1_8

Fig. 8.1 Satellite image of the Yangtze River and the two lakes and their tributaries

complicated river–lake relation separately. Therefore, the river–lake relation can be defined as the interaction between connected rivers and lakes, including exchange of water, natural evolution process of riverbed and lake basin, and the exchange of materials and energy because of the evolution. The river–lake relation of the middle reach of the Yangtze River refers in particular to the interactions between the Yangtze River and the Dongting Lake and the Poyang Lake, which are connected with the Yangtze River.

The connotation of the relation between the Yangtze River and the Dongting Lake means not only the traditional relation of discharge and retention, namely the discharge capacity of Jingjiang River section, water and sediment discharge capacity of the three outlets of Jingjiang River section, the flood retention capacity of Dongting Lake, and the discharge capacity of Chenglingji-Hankou river section, but also the effects on water resources, water environment and water ecology of erosion and deposition evolution and river regime change arising from water, substance and energy exchange between the Jingjiang River and the Dongting Lake. While the relationship between the Yangtze River and the Poyang Lake includes not only the flood retention capacity of the Poyang Lake, but also the effects on water resources, water environment and water ecology of erosion and deposition evolution and river regime change due to the exchange of water, substance and energy.

The river–lake relation has been in constant change since its forming, having a profound effect on the governance and protection of the middle and lower reaches of the Yangtze River. The contributing factors of the river–lake relation mainly include natural evolution process and human activities. Natural factors refer to boundary condition and law of flow-sediment movement, for instance, flow-sediment variation and geological structure. Human activities include damming, reclamation, sand excavation, regulation, and operation of hydraulic projects, etc. The Two Lakes Basin formed before Song Dynasty when the river–lake relation was mainly affected by natural factors. With continuously intensive human activities after Song Dynasty, the relation has been shaped by joint impact of both natural factors and human activities. In particular, the factor of human activities has played a more important role in the relation in modern times. Therefore, coordination of natural factors and human activities becomes the key to building a harmonious river–lake relation.

8.1.2 Major Functions and Roles of the Two Lakes

The Dongting Lake and Poyang Lake, as important parts of the river system and ecological system of the Yangtze River, play an irreplaceable role in the aspects of regulating floods, securing water supply, conserving water resources, maintaining ecological functions and supporting sustainable socio-economic development. They are the support to national important regional development strategies, for instance, the Yangtze River Economic Belt, the Dongting Lake Ecological Economy Zone and the Poyang Lake Ecological Economy Zone, etc.

First, the Two Lakes are important regulation and retention areas for Yangtze floods. Their huge storage capacity is capable of effectively reducing flood peak discharge of the mainstream of the Yangtze River, whilst also capable of storing floods coming from four tributaries of Dongting Lake and five tributaries of Poyang Lake, which greatly reduces the flood control pressure for the middle and lower reaches of the Yangtze River. The maximum annual average inflow into the Dongting Lake is 36,800 m^3/s, while outflow is 27,400 m^3/s, which makes a flood peak discharge reduction rate of 26%. In the case of 1954 and 1998 catastrophic flood events, the three outlets of Jingjiang River section discharged 41% and 27% of the maximum peak discharge of Zhicheng station respectively. The maximum annual average inflow to Poyang Lake is 30,900 m^3/s, while outflow is 8600 m^3/s, which makes a peak discharge reduction rate of 72%. Meanwhile, 28 retention basins are set up in the Two Lakes areas, capable of storing flood of 19 billion m^3. Once a catastrophic flood event occurs, those retention basins can be used to reduce flood peak discharge to ensure the safety of the middle and lower reaches of the Yangtze River.

Second, the Two Lakes serve as important water sources for living and production. They are rich in water resources, with annual average precipitation of about 1100–1800 mm. The annual average water resources volume in the Two Lakes Basin is 361.9 billion m^3, of which the lake areas take up 42.3 billion m^3. Abundant water resources provide important conditions for regional socio-economic development and eco-environment. The lake areas involve 42 counties and cities (including 6 cities in Changsha-Xiangtan-Zhuzhou City Group, Jingzhou in Hubei Province, Nanchang, Jiujiang, Shangrao, Yichun, and Jindezhen in Jiangxi Province), with a population of 256.5 million, farmlands of 1.53 million hm^2, urbanization rate of 55%, producing GDP of 1.77 trillion Yuan. They are also important national agricultural production base, industrial manufacturing base in China, and one of the important growth poles of the Rise of Central China, having a crucial position in the Yangtze Economic Belt Development. The water resources and socio-economic development index of the Two Lakes Basin and the lake areas are as in Tables 8.1 and 8.2.

Table 8.1 Table of the water resources and socio-economic development index of the two lakes basin

	Catchment area (km^2)	Water resources (billion m^3)	Ground water resources (billion m^3)	Population (million)	Farmland (hm^2)	GDP (billion yuan)
Dongting Lake	263,300	208.6	49.32	69.02	4,044,667	3295.1
Poyang Lake	162,200	153.3	37.26	42.83	2,723,333	1705.6
Total	425,500	361.9	86.58	111.85	6,768,000	5000.7

Table 8.2 Table of the water resources and socio-economic development index of the lake region

	Catchment area (km^2)	Water resources (billion m^3)	Ground water resources (billion m^3)	Population (million)	Farmland (hm^2)	GDP (billion yuan)
Dongting Lake	20,100	18.9	4.81	12.750	832,000	1109.4
Poyang Lake	26,300	23.4	4.78	12.90	700,667	661.1
Total	46,400	42.3	9.59	25.65	1,532,667	1770.5

Third, the Two Lakes have vital ecological functions. They, like two kidneys of the Yangtze River, possessing integral ecological service function, are the important parts of the Yangtze River ecological system, and the important ecological functional zones. They are also rich in biodiversity, serving as important international wetlands and habitats for rare migratory birds and breeding habitats for aquatic organisms. The Two Lakes areas include 4 national natural reserves (including east Dongting Lake, west Dongting lake, Poyang Lake and Nanjishan), 8 provincial natural reserves and 27 national aquatic genetic source protection zones and have a total wetland area of 1157 km^2. There are 119 species of fish, 311 species of advanced aquatic plants and 216 species of birds in Dongting Lake. While there are 160 species of fish, 350 species of advanced aquatic plants and 310 species of birds in the Poyang Lake. The Two Lakes are also home to first class national protected animals, such as Chinese sturgeons and while sturgeons, second class national protected animals, such as Yangtze finless porpoise, mullet, etc., and 10 species of state-level endangered birds.

Fourth, the Two Lakes are important waterways connecting upper and lower reaches of the Yangtze River. The features of developed river system, connectivity to the Yangtze River and favorable geographic location make the Two Lakes a convenient inland navigation network, which has played an important role in the transportation between the upper and lower reaches for bulk cargo and containers. The waterways of the Two Lakes are important parts of comprehensive regional transportation system, and the key nodes of the Yangtze golden waterway. The Dongting Lake area has a total navigation mileage of 3609 km^2, including Xiangjiang floodway, Kaihu waterway, Songhu-Lizi waterway, and Lishui floodway, and with two major ports located at Changsha and Yueyang. The Poyang Lake has a total mileage of 734 km^2, including west and east waterways, and Piaosan waterway, with major ports at Jiujiang and Nanchang.

Fifth, the Two Lakes are important scenic spots of Beautiful China. They are very famous for their beautiful landscapes, profound culture heritage and historic relics. Vast waters, mists and ripples, dense water meadows, ganging migrant birds, and beautiful mountains picture a unique landscape of mountains and lakes. The long history in culture left behind a great deal of popular handed-down rhesis and masterpieces, and cultural relics and historic sites, for instance, Yueyang Tower, Du Fu Tomb, Qu Zi Shrine, Bailudong Academy, Xunyang Tower, Tengwang Pavilion. The Two Lakes areas also have many tourist attractions, among which Lushan Mountain Scenic spot is included in the UNESCO's World Cultural Heritage List, and Yueyang Tower, the Peach Garden, and Tengwang Pavilion are included in the important national scenic and culture spots.

8.2 Changes of the River–Lake Relation

8.2.1 River–Lake Relation Before the Construction of the Three Gorges Project

The relationship between the Yangtze River and the Dongting Lake experienced severe changes before the construction of the Three Gorges Project. The Yangtze River and the Dongting Lake were in the status of natural flow over two thousand years ago. Sediments substantially silted up the Yunmeng Pool and Dongting Lake area, which naturally formed embankments along the Yangtze River. Furthermore, with the impact of reclamation and banking, the diversion outlets along both the south and north banks of Jingjiang River Section became smaller and smaller. Up to the year of 1542, embankments along the north bank connected to each other and the diversion outlets on the north bank disappeared with only the two diversion outlets of Taiping and Tiaoxun left on the south bank. Later, the two catastrophic flood events in 1860 and 1870 broke the embankments in the south bank and formed another two outlets of Ouchi and Songzi successively. Therefore, four outlets in total emerged along Jingjiang River section to divert water and sediment to Dongting Lake, until then the pattern of modern river–lake relation forms. The evolution process of the river–lake relation before the construction of the Three Gorges Project can be divided into two stages.

The first stage spans from the time when the modern river lake pattern started taking shape to the 1930s. During this period, the four outlets on Jingjiang River Section diverted huge amount of water and sediments to the Dongting Lake, which resulted in chute cutoff, aggradation and natural cut-off on lower Jingjiang River, and substantial siltation in the Dongting Lake. These changes were beneficial to flood control for the middle and lower reaches of the Yangtze River, but unfavorable for the Dongting Lake.

The second stage is from 1940s until prior to the completion of the Three Gorges Dam. The four outlets diverted less water and sediments (four outlets became three after the Tiaoxuan was blocked in 1958), which resulted in siltation year by year in the four outlet channels. Moreover, the bend of the lower Jingjiang River was cut off, Jingjiang River Section had been eroded, and the river section between Chenglingji and Hankou had been silted up. The flood retention capacity of Dongting Lake was reduced due to the decreased diversion capacity of the three outlets resulting in aggradation. In the meantime, the peak discharge on the river section downstream from Jianli in the Yangtze River increased, which further pushed up the flood of Dongting Lake. As a consequence, the water level at Chenglingji rose. The above changes during this period were bad for the flood control of the Yangtze River and the Dongting Lake. According to statistical analysis of measured runoff data from 1956 and 2002, the average ratio of water diversion of the three outlets was 29.4% from 1956 to1966, while for sediment diversion was 35.4%. Up to the period between 1981 and 2002, the ratio of water diversion was decreased to about 15%, and the ratio of sediment diversion was reduced to 18%. The ratio of water diversion decreased significantly in dry season. The changes of the river–lake relation in the second stage showed various characteristics, such as, the degradation of the river channel of Jingjiang section, the drop of the low water level on Jingjiang section, aggradation and low diversion capacity of the three outlet channels, reduced flood retention capacity of Dongting Lake, longer duration of river dried up on the downstream of the three outlets in the dry season.

The conflict of fighting for land between human and water became prominent before 1980s. The area of Dongting Lake decreased from 4350 km^2 in 1950s down to 2625 km^2 nowadays due to serious siltation and farming, and the area of the Poyang Lake from 5000 to 3708 km^2, which greatly weaken the regulation capacity of the Two Lakes for floods.

8.2.2 Changes of the River–Lake Relation After the Operation of the Three Gorges Project

The dual impacts of both natural changes and human activities, especially construction and operation of reservoirs on the mainstream and major tributaries of the Yangtze River, have led to a series of changes in the runoff and sediment transportation of the Yangtze River. Water and sediment flow into the river–lake system on the middle and lower reaches of the Yangtze River has changed accordingly, in particular the sediment inflow situation, which has resulted in new changes and characteristics in water and sediments transportation, erosion and siltation evolution and river–lake relation.

8.2.2.1 Changes of the Relationship Between the Yangtze River and the Dongting Lake

New changes have been presented in the river–lake relation after the operation of the Three Gorges project besides the changes in the second stage. There are two major reasons, one is the runoff process has changed due to the regulation and storage of the Three Gorges Reservoir, the other is the sediment discharge from the Three Gorges Reservoir has reduced, with the gradation of grain becoming finer. These external factors have taken effect through the internal connection of various aspects resulting in the river–lake relation changed, which has led to further reduction of diversion capacity of the four outlets, further extension of the river dried up period, further decreasing of river lake connectivity. Thus, some problems of water resources and water ecosystem have emerged gradually, and hidden dangers for the flood control of the Yangtze main stream also existed.

The process of water coming in the mainstream of the Yangtze River has changed significantly, and the sediment inflow has significantly reduced. Before the impoundment of the Three Gorges Reservoir, annual average runoff at Yichang, Shashi and Hankou stations downstream of the Three Gorges Reservoir was 436.9 billion m^3, 394.2 billion m^3 and 711.1 billion m^3 respectively. After the impoundment, major stations saw 4–7% less runoff, with an exception at Jianli station due to impact of decreased diversion capacity of the three outlets, where the runoff increased 3%. Before the impoundment of the Three Gorges Reservoir, annual average sediment discharge was 4.92 t, 4.34 t, and 3.98 t respectively at Yichang, Shashi and Hankou stations. After the impoundment, the sediment discharge has decreased gradually along each station, with a decreasing rate ranging from 75 to 93%. The average daily hydrograph at Yichang station before and after the impoundment of the Three Gorges Reservoir is shown in Fig. 8.2.

Jingjiang reach has experienced continuous intensive erosion and fall of water level. During the period between October 2002 and November 2017 when the Three Gorges Reservoir was in operation, the flat riverbed of Jingjiang reach was seriously eroded, with an accumulative erosion amount of 1.051 billion m^3, and average annual amount of 70 million m^3, of which upper Jingjiang river accounted for 60%, and lower Jingjiang river accounted for 40%. Distribution of the erosion and deposition volume along the river section showed that Zhijiang, Shashi, Gongan, Shishou and Jianli river sections accounted for 21%, 24%, 15%, 22% and 18% receptively of the total on the Jingjiang reach. Shashi section riverbed saw the largest annual average erosion volume of 325,000 m^3/km.a, Zhijiang section was the second largest of 254,000 m^3/km.a. While erosion intensity was decreased to some extent at the flat

Fig. 8.2 The average daily hydrograph at Yichang Station before and after the impoundment of the three gorges reservoir

riverbeds of Zhijing and Shishou river sections. With intensive erosion in the mainstream, major stations at Jingjiang reach, such as Zhicheng, Shashi and Luoshan stations, saw obvious drop of water level in dry season with the same flow. During the period between 2003 and 2017, the water level at Zhicheng station decreased by 0.56 m accumulatively when its flow was 7000 m^3/s and decreased by 0.8 m accumulatively when its flow was 10,000 m^3/s.

The river section between Chenglingji and Hankou has turned from disposition to erosion. After the impoundment of the Three Gorges Reservoir, due to the substantial reduction of sediment discharge from the reservoir, it was difficult to replenish the scour sediment on the Jingjiang reach, not reaching the standard of sediment carrying capacity of the river flow on Jingjiang reach. The river section between Chengjingji and Hankou showed erosion or deposition in different years, but in general erosion. During the period between October 2001 and November 2017, total erosion volume of the flat riverbed of this section was 392 million m^3, among which erosion volume of the riverbed below low water level in dry season was 369 m^3 million, accounting for 94%. The riverbed above the low water level in dry season showed slightly erosion.

Water and sediments diverted through the three outlets have continuously decreased, while sediment discharge ratio marginally increased. During the period from 2003 to 2017, the annual average water discharge of the three outlets was 47.99 m^3 billion, 14.54 billion m^3 less than that from 1999 to 2002, with a reduction rate of 23.3%. Corresponding water diversion ratio dropped from 14% during the period between 1999 and 2002 down to 11.6% between 2003 and 2017. During the period from 2003 to 2017, the annual average sediment discharge was 8,670,000 tons, 48,030,000 tons less than that from 1999 to 2002, with a reduction rate of 84.7%. However, the decrease of sediment discharge was favorable to the relief of aggradation in the Dongting Lake.

The channels of the three outlets have turned into degradation in general, due to low sediment concentration on Jingjiang reach since the operation of the Three Gorges Reservoir according to statistics. During the period from 2003 to 2017, total volume of erosion was 172 million m^3, of which Songzi River was 100 million m^3, accounting for 58.2% of the total, Hudu River was 220 million m^3, accounting for 12.7%, Songhu floodway was 7 million m^3, accounting for 7.9%, and Ouchi River was 36 million m^3, accounting for 21.2%.

Aggradation in the Dongting Lake has obviously slowed down. During the period from 1995 to 2003 before the impoundment of the Three Gorges Reservoir, aggradation dominated in the Dongting Lake. Muping Lake in West Dongting Lake, East to Yangliu Pool in South Dongting Lake and the East Dongting Lake were in the status of aggradation. While after impoundment, sediment inflow into Dongting Lake dramatically dropped. Comparing to the average from 1996 to 2002 before impoundment, the sediment discharge from the three outlets and the four tributaries decreased by 86.8% and 47.1% respectively from 2003 to 2016. As a result, aggradation and deposition rate in the lake area reveal a distinct trend of declining, with total volume of sediment deposition dropped down to −2,110,000 tons, showing a change from deposition to erosion. According to the observed topography data of 1995, 2003 and 2011 from the Bureau of Hydrology of CWRC, small volume of deposition mainly occurred in the west part of South Dongting Lake and the south part of East Dongting Lake, and the average degradation depth in the lake area was 10.9 m, among which the East Dongting Lake had the biggest degradation depth of 19 m.

The dry season for Dongting Lake arrives earlier than usual, with longer duration. Under the influence of impoundment of the Three Gorges Reservoir and other key reservoirs in post-flood season and erosion in the mainstream, the water level of the Yangtze mainstream has dropped during September and October, which has led to water being drawn out from Dongting Lake quickly. As a consequence, the water level of Dongting Lake becomes lower in the dry season. The average water level of Dongting Lake was dropped by 0.36–1.82 m during the period from 2003 to 2017 comparing to that from 1980 to 2002.

The daily average hydrograph at Chenglingji station (Lianhua Pool) before and after the construction of the Three Gorges Project is shown in Fig. 8.3. The change of volume and rate of deposition in the Dongting Lake area from 1956 to 2016 is shown in Fig. 8.4.

8.2.2.2 Changes of the Relationship Between the Yangtze River and the Poyang Lake

The dry season of the Poyang Lake arrives earlier than usual with longer duration. Under the influence of impoundment of the Three Gorges Reservoir and other key reservoirs in post-flood season and aggregation, the water level of the Yangtze mainstream has dropped during September and October, which has led to water drawn out from Poyang Lake quickly. As a consequence, the water level of Poyang Lake becomes lower in dry season. The average water level of Poyang Lake was dropped

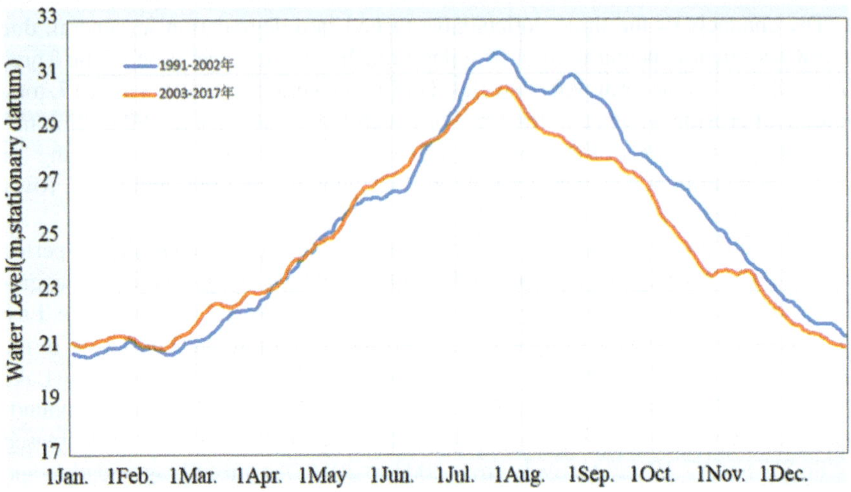

Fig. 8.3 Daily average hydrograph at Chenglingji station before and after the construction of the three gorges project

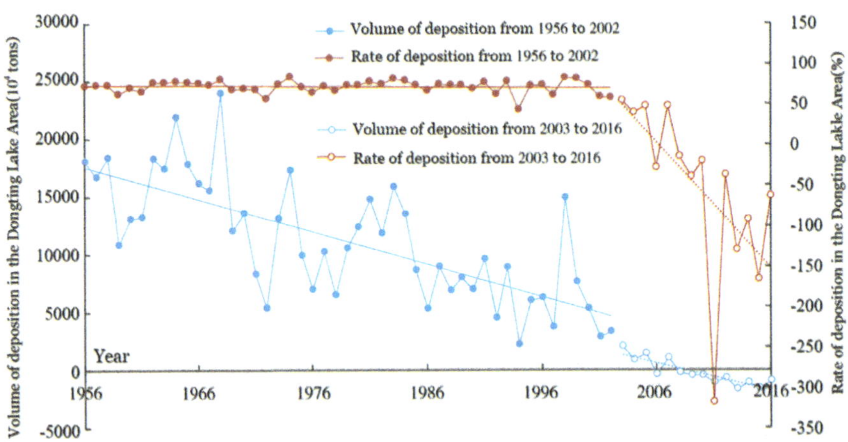

Fig. 8.4 The change of volume and rate of deposition in the Dongting lake area from 1956 to 2016

0.56–2.31 m during the period from 2003 to 2017 comparing to that from 1980 to 2002. The low level appeared about 31 days earlier. The duration of water level below 10 m at Xingzi Station extended 20 days than usual. Xingzi, Kangshan, and Yongshan station witnessed historical lowest water level in records in 2004. The daily average hydrograph at Hukou station before and after the construction of the Three Gorges Dam is shown in Fig. 8.5.

The drop of water level in September is mainly influenced by impoundment of the Three Gorges, while the drop of water level in October is attributed to the joint impact of impoundment of the Three Gorges and other reservoirs and natural runoff

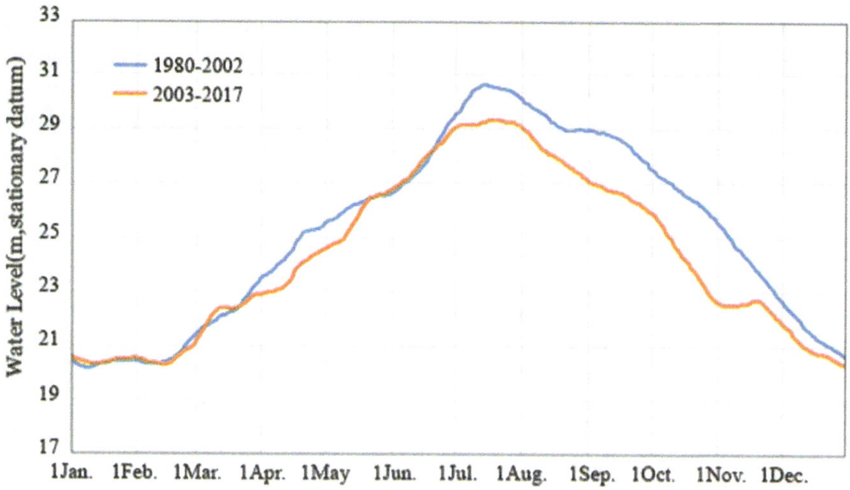

Fig. 8.5 Daily average hydrograph at Hukou station before and after the construction of the Three Gorges Dam

variation, and the drop of water level in November is because of degradation in the Yangtze mainstream, natural runoff variation and impoundment of other reservoirs. In dry season, the discharge of the Three Gorges Reservoir increases, which alleviates the impact of erosion on the lowest level.

Water inflow from the five tributaries into Poyang Lake has no apparent changs, while sediment inflow into the Lake has decreased obviously and sediment outflow has increased. The annual average amount of water and sediment into the Poyang Lake from the five tributaries (not including interval inflow) from 1956 to 2017 were 109.4 billion m^3 and 12,200,000 tons respectively. The runoff in this period showed no obvious trend change, while sediment discharge to the Poyang Lake has showed a continuous decline trend since 1961, and so did the annual sediment discharge from Hukou to the Yangtze River from 1956 to 2002. During the period from 2003 to 2017, the annual average sediment discharge from the five tributaries reduced to 5,820,000 tons and by 43.5% comparing to that from 1991 to 2002. In the meantime, the sediment outflow of the Poyang Lake showed an increase trend by 61.2% comparing with that during the period from 1991 to 2002.

The flow backward to Poyang Lake from the Yangtze River, the maximum backward flow volume, the average annual duration of flowing backward have showed a decline trend. During the 54 years from 1956 to 2017, 764 days had backward flow into the Poyang Lake, with total volume of 145.5 billion m^3, and average volume of 2.85 billion m^3. The biggest backward flow was 6160 m^3/s on Sep 9, 2003.

The river channel connecting to the Poyang Lake and the Yangtze River has experienced severe down cutting, and the water level at Xizing station has decreased obviously. According to the analysis on the sediment inflow and outflow in recent years, and analysis on the observed topography data of the lake area in 1998 and

2010, it was found that the Poyang Lake area was in slow aggradation status before 2003. During the period from 1998 to 2010, affected by backward erosion resulting from the dropping water level on the mainstream as well as the sand mining, it was shown that the river channel connecting to the Poyang Lake and the Yangtze River experienced dramatic down cutting, the riverbed elevation in lowest water level was in the trend of decrease, and the area of cross sections below the water level of 15 m became obviously bigger. The average daily water level difference between Xingzi and Hukou stations during the period from 1956 to 2002 was 0.57 m, while it became 0.8 m when the water level at Hukou station was below 12 m. While the difference was 0.28 m due to the down cutting effect during the period from 2003 to 2012, and it became 0.35 m when the water level at Hukou station was below 12 m. Comparing the two periods, it was found that the water level difference became smaller, and the water level of Xingzi station reduced visibly when Hukou station remained the same level during different periods.

8.2.2.3 Variation Trend of the River–Lake Relation

With the successive construction of several key reservoirs in the future, including Wudongde, Baihetan, Shuangjiangkou, and Lianghekou, further evolution will take place in the river–lake relation. Under the influence of the Three Gorges and other major key reservoirs on the upper stream, further long-distance degradation will occur in the main channel along the middle and lower reaches of the Yangtze River. Meanwhile, impoundment of upper stream reservoirs and degradation on the mainstream channel will further affect the hydrological regime of the Two Lakes.

8.3 Existing Problems in Rehabilitation and Protection of the Two Lakes

Under the long-term influence of natural geographic condition, human activities and climate change, the river–lake relation has experienced continuous evolution. Especially major key reservoirs constructed in succession on the mainstream and tributaries of the Yangtze River have exerted tremendous and comprehensive benefits in the aspects of flood control, power generation and navigation. In the meantime, the river–lake relation has experienced new changes, which places a long-term influence on the eco-environment, urban and rural water supply, and flood control to the middle and lower reaches of the Yangtze River.

First, efforts on ecological protection should be further enhanced. Due to the change of the river–lake relation, the water surface area of the Two Lakes in dry season dramatically has shrunk and reduced, only accounting for 5–15% of the total lake area. The four outlets river system and the branches of the estuarial part of Ganjiang River had less water and longer river dried up duration, the floodplains of the Two

Lakes have been exposed ahead of time, the landscape of wetland has been destroyed in fragmentation, and the ecosystem function has declined. Furthermore, on account of overfishing, reclamation, pollution and construction activities in the lake area, the habit has been destroyed to certain extent, with degraded biodiversity. Since 1970s, the fish species in the Two Lakes area have reduced by a third, and fishery production has reduced by half. The number of birds in the Dongting Lake reduced from around 500,000 to less than 100,000, while that in the Poyang Lake reduced from around 700,000 to around 200,000. Rare animals such as finless porpoise, Chinese sturgeon, paddlefish, etc., are becoming more endangered. In recent years, efforts have been made to protect the ecosystem of the Two Lakes, and progresses have been achieved. For instance, the number of wintering birds of Dongting Lake increased from 60,000 in 2007 to 180,000 in 2013, while for the Poyang Lake increased from 300,000 in 2009 to 700,000 in 2015. However, the protection of rare animals such as the finless porpoise hasn't shown an obvious improvement. Therefore, further enhancement on protection and rehabilitation of the eco-environment of the Two Lakes is still needed.

Second, there is heavy responsibility and a long way to go for water environment pollution control. The sources of pollution of the Two Lakes are very complicated, including agriculture diffuse pollution represented by livestock breeding and fertilizers, and point pollution represented by domestic sewage and industrial waste. The COD discharge into the Dongting Lake area in 2016 was 143,800 tons, ammonia and nitrogen discharge were 21,000 tons. While the COD discharge into the Poyang Lake area was 304,600 tons, ammonia and nitrogen discharge were 34,000 tons. The pollutant discharge volume into the Two Lakes area has already exceeded the limitation of their pollutant assimilation capacity, which threatens the safety of water environment in the lake area. Efforts on pollution control in recent years in the lake area have been strengthened; as a result, the water environment has shown improvement in general. The EI values of nutrition state index of the Two Lakes were reduced from 60 and 56 in 2016 down to 55 and 48 in 2017. However, the water environment pollution situation is still severe in the lake area as the socio-economic development puts forward higher requirement on water environment rehabilitation. Rehabilitation of water environment is still an arduous task.

Third, urban and rural water supply safety is under threat. The Two Lakes are in serious seasonal water shortage, which increases the difficulties of water use for industry and living, with affected farmland area of 292,667 hm^2, and affected population of 4,310,000. Water shortage induced by water quality is very common in the lakes area, and algae bloom often occurs due to decreased self-purifying capacity of the lake. Especially in the four outlets area, iron and manganese concentration have reached a hazardous level, which threatens the drinking water safety of local inhabitants. The existing water source projects in the lake area have a low regulation capacity, and emergency stand-by water sources are in shortage. The existing irrigation projects are aging, with channels blocked and structures damaged, and furthermore, related field works are not yet in place, leading to serious leakage and a low utilization efficiency of irrigation water. The pipeline network for urban water supply is aging as well, with a leakage rate exceeding 16%. Water use efficiency in the Two Lakes area is not high. Water saving irrigation area is only 140,000 hm^2

at present, accounting for only 12% of effective irrigation area. Effective utilization coefficient of irrigation water is about 0.49. Ten-thousand-yuan industrial added value water consumption are 70 m^3 and 65 m^3 respectively in the Dongting Lake and the Poyang Lake, higher than national average. Reuse rate of industrial water is only 60%.

Fourth, there are still areas of weakness in flood control and disaster mitigation system. The extraordinary flood events occurred in 2016 and 2017 respectively in the Two Lakes area exposed the weak links. Embankments along the Two Lakes have not yet been fully met the standards, and hidden dangers still existed in the embankments of important polders. Safety construction for relevant retention basins has lagged behind, and flood diversion works of most retention basins are yet to be improved, which makes diversion even harder. Due to the back-up effect of high waters in stream, a large volume of water logged in some of detention basins cannot be out drained by gravity. On the other hand, the pumping facilities within the polders are aging, which resulted in low drainage capacity for waterlogging. The capacity of flood control of medium and small rivers in the lake area is relatively poor, which makes the regulation an arduous task. The construction of flood forecast and warning system at local level has lagged behind, and information platform is incomplete.

Fifth, the integrated management of the lake region needs to be strengthened urgently. Multi-industries and multi-sectors will be involved when it comes to the rehabilitation and protection of the Two Lakes. For a long time, effective coordination mechanism is not in place among related regions, responsibilities across different sectors overlap, and lake management system and mechanism have yet to be straightened out. Awareness of lake protection is poor in some places, main responsibilities of local authorities for lake protection have not been fully implemented, and accountability system for trans-provincial water resources protection is yet to be improved. The laws and regulations on lake protection are not perfect, because the illegal cost is low, the water administrative law enforcement ability is weak, the traditional thinking of attaching more importance on economic development over ecological protection, and polluting first and treatment later has not been eliminated, illegal sand excavation, illegal sewage, encroaching on the waters, destruction of ecology and other problems occur from time to time. The monitoring and supervision system for the water resources in the lake area is not improved, and the data from relevant industry monitoring stations is lack of integration and sharing.

8.4 Strategies for Rehabilitation and Protection of the Two Lakes

8.4.1 Overall Protection and Restoration of the Ecological Environment of the Two Lakes

Efforts should be made to strengthen management and control of the ecological conservation redline and develop the access system that is appropriate to the control of the Redline. Ecological river network connecting projects for the Two Lakes should be implemented orderly so as to improve the river system connectivity and the water circulation dynamics condition and enhance restoration capacity of the water ecosystem. Overall regulation of blocked rivers and ponds in the polders of the Two Lakes should be carried out to restore natural hydraulic links and reduce adverse impact of blocked rivers and lakes on biodiversity. Researches should be carried out on the restoration of fish migration passage for existing hydraulic projects on the rivers flowing into the lakes, aiming to gradually restore aquatic organism passages for the lake area and major tributaries.

We should adhere to natural restoration and give priority to restoration and comprehensively regulation of natural wetlands with fragmented habitat and severe function decline. Furthermore, we should also implement the strategy of "returning farmlands to lakes and wetlands", consolidate the outcomes of "removing river-encroaching polders to give room back to floods and returning reclaimed lands to lakes", and promote protection and restoration of mountains, rivers, forests, farmlands, lakes and grasslands.

Efforts should be made to strengthen protection of the sources area for drinking water in the lake area. Agricultural non-point source pollution control projects should be implemented in 16 drinking water sources areas including the water source area of Tieshan Reservoir in the Dongting Lake area.

The prevention and control of agricultural pollution from internal and non-point sources should be intensified. Ecological dredging should be implemented with application of eco-friendly technologies for dredging bottom sediment to reduce internal-source load and pollution risk, so that the purifying ability of bottom sediment will be improved. Efforts on ecological measures to prevent nitrogen and phosphorus loss from farmland should also be made. Ecological and high-tech agriculture need to be rolled out to effectively reduce the use and loss of pesticides and fertilizers, such as fertilization technology based on soil measuring, and biological pest control technology. For those irrigation districts where recession water pollution is serious, ecological ditches and rainwater storage & utilization works must be built to separate irrigation and drainage, strengthen interception of nitrogen and phosphorus nutrients discharge from farmlands, and reduce the amount of pollutants entering rivers.

The habitats for river finless porpoises, Chinese sturgeons, Chinese mergansers, white storks, white-headed cranes and other rare animals in the Two Lakes should be strictly protected. Ecological measures should be taken to restore and rebuild destroyed habitats for important migratory bird and important spawning areas and migration passages of rare fish. Efforts on protection of nature reserves and aquatic germplasm resources reserves should be strengthened. A complete fishing ban should be carried out, extend fishing ban season, and expand fishing ban area. In the meantime, forbidden fishing gears will be strictly managed and banned in accordance with relevant laws, fishmen will be encouraged to transfer to other industries, and new industries such as tourism and game fishing will be supported and developed. Comprehensive tracing monitoring of the ecological environment of the Two Lakes should be intensified, and necessary research on key issues will be carried out, such as maintenance of habitats for birds and fishes. Moreover, practice on ecological regulation should be enhanced.

Ecological protection compensation mechanism and related scientific compensation standard system should be studied and established in order to provide a quantitative basis for ecological compensation. Financial transfer payment system and eco-friendly environmental taxation system should be improved or established, and market-based ecological compensation methods should be promoted.

8.4.2 Full Implementations of Water Conservation and Optimization of Water Resources Allocation

Development of a water-saving society should be sped up to improve guarantee capacity of water supply. The strictest possible systems for water resource management, the "Three Red Lines" of the total quantity of water resources exploitation and utilization, water use efficiency, and the total amount of limit pollutant assimilation should be strictly implemented. Actions on dual control of the total amount and intensity of resources consumption will be carried out. By 2025 and 2035, the total quantity of water use in Poyang Lake and Dongting Lake will be limited to 16.313 billion m^3 and 16.771 billion m^3 respectively. Measures will be taken to improve the capability of effective water supply for the Two Lakes area, for instance, construction of water source projects, implementation of comprehensive regulation projects for the four outlets river system of the Dongting Lake, and upgrade and construction of storage, diversion and pumping works. Furthermore, construction of urban and rural water supply facilities should be strengthened to improve regional and networked water supply by taking measures of construction of water source works and affiliated works, and upgrade, networking, and industrialization of water supply facilities. The drinking water safety of rural areas needs to be further improved. For areas where groundwater quality is not up to standard, surface water sources should be utilized for drinking water if condition allows. Extension and supporting works for irrigation districts will be constructed to improve the irrigation and drainage system of farmlands, create better irrigation conditions, and improve water supply capacity, ensuring probability of irrigation water and irrigation efficiency.

Emergency water supply safeguards should be improved. In order to cope with drought event and water pollution emergencies, governments at all levels in the lake area should strengthen the construction of urban water supply emergency response system and emergency rescue mechanism and formulate contingency water supply plans for their respective jurisdictional regions above the county level. For those cities where surface water is the main source for drinking water, emergency response classification and plan should be made respectively based on natural and human factors, such as drought disasters, water source pollution, earthquakes, engineering accidents, and emergency events, to enable a smooth implementation of emergency water dispatch, ensure disaster losses under control, minimize adverse impact of disasters on local economic and social life, and secure the safety of water supply to the utmost extent. For those cities above the prefectural level with only one single water source, alternate or emergency water source should be in place before 2025. County-level cities should regard surrounding reservoirs and lakes as alternative emergency water sources or expand existing water supply works.

8.4.3 Improvement of Flood Control, Drainage and Disaster Mitigation System for the Two Lakes Areas

Embankments should be reinforced and upgraded to reach standard. Important embankments of the polders in the Two Lakes areas need to be upgraded based on their conditions and the dangers occurred in recent years.

Construction for the safety of flood storage and detention basins should be enhanced. Construction to improve safety should be carried out in the important national flood storage and detention basins, such as Qianliang Lake, Gongshuangcha, east Datong polder, Minzhu, Chengxi, Jianshe, Honghudong, Kangshan, etc., with focus on Qianliang Lake, Gongshuangcha, east Datong polder. The revision of the layout of the flood storage and detention basins along the Yangtze River should be studied and verified accordingly.

Construction of drainage works in waterlogging-prone areas should be strengthened. Implementation Plan of Accelerating Post-Disaster Construction of Water Conservancy Weak Links should be used to overall coordinate the relation between basin flood control and regional drainage. On the basis of protecting rivers, lakes, low-lying land and other detention bodies, and under the condition of maintaining water surface area ratio, measures, such as dredging channels, setting up and upgrading drainage pumping stations, should be taken to increase the drainage by gravity and pumping-drainage capacity for waterlogging-prone areas. Construction should focus on external drainage pumping stations and main drainage channels, and regulation of the canal systems, embankments of lakes (polders) and culvert gates within waterlogging areas will be properly considered.

Regulation of the lake area and the river estuary floodways should be strengthened. Regulation of floodway should be conducted to expand cross-sections and increase the flood discharge floodway is to expand cross-section and increase the flood discharge capacity. About 600 km-long floodways in the Dongting Lake basin area (among them, about 470 km in the four outlet river system and 130 km in the lake area) and 279 km-long floodways in the Poyang Lake basin should be regulated to increase the flood discharge capacity and reduce siltation in the channels by measures of dredging, clearing, widening and desilting.

Construction of reservoirs on major tributaries of the Two Lakes should be strengthened. The reservoirs on the major tributaries are in charge of their own flood control; meanwhile, they can reduce the peak discharge and the water level of the lake when the Yangtze River flood occurs at the same time, which is proved to be one of the effective measures to solve the flood problems of the Two Lakes area. Meanwhile, efforts will be made to expand the storage capacity of Wuqiangxi Reservoir on the Yuanjiang River, and promote the completion of planned reservoirs, so as to give full play to their roles in flood detention and peak cutting.

Regulation of small and medium-sized rivers should be accelerated. Priorities will be given to those river sections experienced severe flood and waterlogging disasters in recent years and with densely populated counties, towns and villages. Major measures include embankment and revetment reinforcement and construction, river channel dredging, and drainage project construction, etc. Attention should be paid to protect the water ecological environment in the meantime.

Non-structural measures for flood control should be further strengthened. The meteorological and hydrological station network needs to be enhanced to improve the monitoring, warning and forecasting level of hazardous weather such as storms. Construction of flood control commanding system should be improved, in particular the construction of the flood warning and forecasting system at the grass-roots level. Publicity and education for flood control knowledge and emergency response drilling must be strengthened to raise public's awareness of flood control, disaster mitigation, and self and mutual rescue abilities. Flood detention basins, floodplains and polders must be well managed to ensure flood diversion and detention (discharge) in case of extraordinary floods. Studies on joint operation of river basin flood control system and flood control operation for flood exceeding the designed level of the lake area should be carried out. Flood insurance is highly encouraged and supported.

8.4.4 Implementation of Key Projects to Optimize the River–Lake Relationship

Based on studying the changes and evolution trend of the river–lake relationship, we should strengthen the joint operation of cascade reservoirs on the upper and middle reaches, promote the comprehensive regulation projects of the Two Lakes in an orderly manner, build a harmonious river–lake relation, ensure the ecological base flow and ecological environment water demand of the rivers flowing into the lake, the ecological water level and the ecological area of the lake, and improve the carrying capacity of water resources and the carrying capacity of the water ecological environment of the Two Lakes, so as to achieve the ideal of "win–win for both rivers and lakes, ecological priority and multi-functions".

Joint operation of the Three Gorges Reservoir and the other key reservoirs on the upper and middle reaches of the Yangtze River and the reservoirs on the tributaries of the Two Lakes should be implemented. In line with the principle of safety first, controllable risk, and maximum benefit, and the principle of other benefits subordinated to flood control, regions subordinated to the whole basin, and electricity operation subordinated to flood control operation, efforts should be made to alleviate the impact brought by the decrease of water level in dry season of the Two Lakes by scientifically formulating flood operation plan, reasonably scheduling for reservoir storage and discharge, and making an overall consideration to ecological safety.

Aiming at problems, such as the reduction of diversion of the four outlets of the Dongting Lake, aggradation in river channels, the poor discharge, and prominent flood control problems of Songli area, efforts should be made to strengthen the comprehensive regulation of the four outlets river system, promote the construction of Songzikou sluice gate, restore the connectivity between the rivers and the lakes in dry season, enhance the carrying capacity of water resources and water environment in the four outlets river system, improve conditions of water supply, irrigation, flood control and navigation, and maintain stable and virtuous cycle of the wetland ecosystem of the Dongting Lake. Furthermore, preliminary studies and demonstration on Poyang Lake complex hydro project and Chenglingji complex hydro project should be carried out accordingly.

8.4.5 Enhancement of the Integrated Management of the Two Lakes Areas

Water-related law and regulation system should be established and improved. Laws, regulations and rules should be issued as soon as possible, including the Yangtze River Protection Law, Regulations on Dongting Lake Protection, Regulations on Lake Protection in Jiangxi Province, Regulations on the Flood Detention Basin Management of Dongting Lake area in Hunan Province, Regulations on Rural Water Supply Management in Jiangxi Province, as well as Implementation Rules for Supervision and Management of Water Functional Zones and Implementation Rules for Supervision and Administration of Sewage Outlets into Rivers. Management systems for water resources allocation, conservation and protection, flood control and drought mitigation, rural water conservancy, water and soil conservation, and sewage outlets into rivers should be studied and formulated. Laws and regulations for integrated river basin management should be gradually improved.

The Opinions on Fully Implementation of the River Chief System and the Guidance on the Implementation of the Lake Chief System should be thoroughly implemented. Efforts should be made to establish and improve the Lake Chief System for the Dongting Lake and the Poyang Lake, and implement territorial management responsibilities with governmental leadership responsibility system as the core. Grid management should be applied to ensure that all waters in the Lake area have a clear body of responsibility. Responsibilities of Lake Chief need to be clearly defined, and management and protection of lakes and connecting rivers should be well coordinated. Main tasks of the Lake Chief system should be fully implemented, including strictly controlling the space of lake area, strengthening management and protection of lake shoreline, strengthening lake water resources protection and water pollution prevention, strengthening lake water environment improvement, implementation of ecological governance and restoration, and improvement of the mechanism for law enforcement and supervision. Working mechanism of Lake Chief system should be further improved to ensure effective lake protection, by defining lake management authorities, strengthening coordination across sectors, being strict in accountabilities, and strengthening monitoring and supervision. Roles in coordination, guidance, supervision and monitoring of river basin authorities should be fully played, and a long-term mechanism for river and lake protection featuring integrated management of land and waters, joint management across sectors, and public participation should be established.

Water-related affairs should be better managed. Management of water intake and drainage should be strengthened. Efforts should be made to implement the strictest possible systems for water resources management and its appraisal system, set up the "Three Red Lines" to make water resources exploitation and utilization under control, improve water use efficiency, and limit water resources pollution, and conduct assessment for the implementation of the strictest possible systems for water resources management. Furthermore, shoreline management and protection should be strengthened as well. Work has to be done to define the scope of river and lake management based on laws and to improve the protection and economical utilization of lake shorelines, define protection area, reservation area, limited utilization area and development area for shoreline protection, and strictly manage and control the utilization of shoreline space. Sand excavation in rivers and lakes must be better controlled. Efforts should be made to formulate sand excavation plans for rivers and lakes in a scientific way, define excavation zones and forbidden zones, excavation period and forbidden period, determine annual total amount of sand excavation and the number of sand excavation vessels in excavation areas, and strengthen the supervision and inspection of the implementation of sand excavation plans. Management of flood detention basins should be strengthened. Measures will be taken to normalize the economic and social activities in flood detention basins, step up improved distribution, structural adjustment, limit high-risk economic development activities, strengthen land use management in flood detention basins, and bring population growth under control, so as to make the production and living in the flood detention basins compatible with the requirements of flood control. Law enforcement for water-related affairs should be comprehensively promoted. Efforts should be made to intensify integrated supervision and management for water resources, water environment, water ecology, shoreline, and sand excavation, and take stern action in illegal water-related activities.

Joint operation system for key reservoirs need to be established. The joint operation of major reservoirs on the Yangtze River and the Two Lakes is helpful to mitigate flood disasters in the Two Lakes areas, improve utilization conditions of water resources and protect the water ecological environment. Measures should be taken to establish a consultative mechanism and management system for joint operation, encouraging wide participation of stakeholders, such as improving laws and regulations, and perfecting management system, establishing compensation mechanism, strengthening technical support, intensifying supervision and management, etc. Emergency operation system should be improved to deal with emergencies, such as severe drought, water pollution, water safety accidents and engineering accidents, etc.

Capacity building for monitoring and basic research needs to be strengthened. Efforts should be made to strengthen the construction of a comprehensive monitoring network containing hydrological station network, river and lake monitoring station network, water environment monitoring station network (including stations for sewage outlets on rivers), water ecology monitoring station network and soil erosion monitoring station network, as well as field science experiment research and observation station network, improve monitoring and warning system and realize the sharing of information. Moreover, work has to be done to monitor dynamic changes of the water quality of rivers flowing into lakes and trans-provincial water bodies. It is planned to set up new monitoring stations in related water function zones along the river sections of the tributaries of the lakes, including Songzi River, Hudu River, Ouchi River of the Dongting Lake, and Ganjiang River, Fuhe River, Xinjiang River, Raohe River and Xiushui of the Poyang Lake. Meanwhile, on-line monitoring devices for key sewage outlets on lakes (rivers) will be installed.

In-depth research on the river–lake relation should be carried out, focusing on the evolution mechanism, rules and trend, as well as the impacts of the river–lake relation variation on flood control, water resources, water environment and water ecology and countermeasures.

Efforts should also be made to improve the capacity of scientific and technological innovation in water conservancy. Scientific research platforms in key fields must be set up, for instance, rehabilitation and protection of water eco-environment, water resources allocation, flood control operation, river sediment and reservoir operation. Research, development and innovation on a number of key technologies should be strengthened, such as, the evolution of the river–lake relation and the solutions, eutrophication prevention and control in lakes (reservoirs), water ecological restoration, superposition effect of impacts exerted by key water conservancy projects and hydropower projects on the mainstream of the Yangtze River and major tributaries, joint operation of cascade reservoirs, flood risk management, flood utilization and integrated dispatch of water resources.

Chapter 9
Inter-basin Water Transfer

China's gross amount of water resources is 2841.2 billion m^3, and the per capita amount of water resources is about 2110 m^3, which is only 28% of the world's per capita share. China is not at an advantageous position in terms of water resources endowment, and in addition, the geographical distribution of water resources is very uneven. Water is abundant in the south, but much less in the north, which is not fully compatible with the socio-economic development of the country. The Yangtze River Basin is a region with relatively rich water resources in China. The inter-basin water transfer can alleviate the grave water shortage problem in the northern regions, promote the coordinated development among demography, resources, and socio-economy, holding a strategic standing in optimizing allocation of water resources across China.

9.1 Strategy and Conception

The Yangtze River Basin is endowed with plenty of water resources, with a gross amount of 995.8 billion m^3, of which the gross water availability is 282.7 billion m^3. It is projected that in the target year of 2030, the water consumption in the Basin will reach 131.2 billion m^3, and the volume of water entering the sea will account for more than 80% of the Basin's gross amount of water resources. In addition, the geographical distribution of water resources is very uneven. The Basin's rich water resources make it a key strategic water source feeding water transfers to other parts of China. From the perspective of water quantity, after the local demand for water in the Yangtze River Basin itself is met, a portion of the superfluous water can be diverted to alleviate the water shortages in adjacent river basins. In view of the terrain, topography and engineering conditions, the Yangtze River, from west to east, runs

© Changjiang Press (Wuhan) Co., Ltd. 2022

Changjiang Water Resources Commission of Ministry of Water Resources of the People's Republic of China, *Report on Yangtze River Rehabilitation and Protection 2019*,
https://doi.org/10.1007/978-981-16-4927-1_9

across a vast land territory of the country. Its upper reaches are close to the arid regions in the northwest while the middle and lower reaches are adjacent to the most water-deficient Huang-Huai-Hai Plain and the Jiaodong Region (Eastern Shandong Province). The superior geographical conditions render the launch of inter-basin water transfer projects with notable economic and technological advantages.

The basins of the Yellow River, Huaihe River and Haihe River, lying to the north of the Yangtze River Basin, are rich in land, energy and mineral resources, and have good light and heat conditions. Their population, cultivated land and GDP all account for more than one third of the country's averages, but water resources in these regions are scarce, where the gross amount of water resources only accounts for 7% of the country's total, and the per capita share is less than 450 m^3. It is one of the regions where to find the worst incompatibility between water resources bearing capacity and socio-economic development, and the biggest gap between water supply and demand, featuring severe shortage of water resources. Since the 1980s, the Huang-Huai-Hai Plain has experienced continuous drought, with expanding scope and increasing degree of water shortage. On the one hand, global warming and human activities have, to varying degrees, impacted the quantity, quality and tempo-spatial distribution of water resources. Especially in the arid and semi-arid northern regions, the gross amount of water resources is decreasing. Affected by the monsoon climate, water resources vary greatly from year to year, and the intra-yearly distribution of precipitation and runoff is also uneven, with annual precipitation mainly concentrated in the rainy season, making water resource utilization even harder. In the early 1980s, the water shortage was only limited to parts of Beijing, Tianjin and Hebei Province. After 2000, 43% of the area in the Huang-Huai-Hai basins had a water deficit rate of more than 10%, and the proportion of the area stricken by water shortage in the Haihe River Basin was as high as 70%. The problem of urban water shortage became more and more serious. On the other hand, with the rapid increase of industrial, agricultural and domestic water consumption, water pollution has exacerbated, and water environment deteriorated. In order to meet the demands of water for socio-economic development, the proportion of groundwater use in water deficit areas has been rising, and its over-exploitation has led to a substantial drop in groundwater tables, forming an ever-expanding regional funnel of dropping groundwater table. Constrained by the physical conditions of water resources, it is hardly possible to augment the water supply by exploiting local runoffs in the Huang-Huai-Hai Basins, since over 90% of the local water availability has already been used, and in some districts this figure has even exceeded 100%. Over-exploitation of water resources has brought about a series of eco-environmental problems and aggravated the existing water crisis. Therefore, it is imperative to import water from other river basins. Addressing the water resources shortage in the Huang-Huai-Hai Basin has always been the target of conceiving water transfers from the Yangtze River Basin.

In other developments, the Central Yunnan Plateau lies on the dividing crest of 4 watersheds, namely the Jinsha River, Lancang River, Honghe River and Nanpan River (draining to the Pearl River). This region features low precipitation and short-length streams with small discharges. But this region contains the megalopolis Kunming and other important cities such as Qujing, Yuxi, Chuxiong and Dali, which form the kernel of the Yunnan Province taking such shares as 39.5% of the province's population, 26.2% of the cultivated land and 67.3% of the provincial GDP. By contrast, the per capita amount of water resources is only in the range of from 139 to 560 m^3, in the category of extreme water scarcity. The Guanzhong area (also known as the central Shaanxi region), adjacent to the Yangtze River Basin, is an economically, culturally and politically important area of Shaanxi province, where is a concentration of 64% of the province's population, 56% of cultivated land and 82% of GDP. Whereas, its per capita amount of water resources is only 370 m^3, apparently in a state of extreme water stress.

The regions mentioned above are suffering from the severe shortage of water resources, featuring more and more acute incompatibility of water resource bearing capacity with socio-economy and ecological environment. Self-sufficiency of water resources can hardly be a solution, instead, water transfer from the adjacent Yangtze River Basin is taken as a strategic measure for re-allocation of water resources to alleviate water shortages and improve the ecological environment.

9.2 Planning and Layout

The Key Points of Plan for Comprehensive Utilization of the Yangtze River Basin issued in 1959 proposed a preliminary scheme of water diversion from the upper, middle and lower reaches of the Yangtze River. After more than 30 years of verification and research, the Executive Report on Plan for Comprehensive Utilization of the Yangtze River Basin (1990) further proposed the planning schemes for the East, Middle and West routes of the South-to-North Water Transfer Project and Yangtze-to-Huaihe Water Diversion Project. The Master Plan of the Yangtze River Basin (2012–2030), taking grounds from the call for rational allocation of water resources in the Yangtze River Basin, proposed the East, Middle and West routes of the South-to-North Water Transfer Project, under the principle quoted as "water conservation before water transfer; pollution control before water supply; environmental protection before water use". The Master Plan also emphasized intensified measures for water conservation, pollution control and eco-environmental protection, and amended the transferable water volume and the previous planning scheme of the South-North Water Transfer Project.

The East Route of South-to-North Water Transfer Project is located on the eastern part of the "third-terrace" region. It draws water directly from the downstream of the Yangtze River mainstream where the stream flow is sufficient enough. The water is then conveyed to Tianjin Municipality and Shandong Peninsula passing Jiangsu, Shandong and Hebei provinces, for urban and agricultural water uses along the route. Because of the low ground elevation of the East Route, water conveyance from the intake to the south bank of the Yellow River relies on pumping stations at intervals on the way, while after crossing the Yellow River the water flows by gravity to the eastern parts of the Huang-Huai-Hai Plain and the Jiaodong area.

In near term, the Middle Route project diverts water from the Hanjiang River, a tributary of the Yangtze River; and in the long run, water will be replenished from the Three Gorges Reservoir. The Middle Route canal traverses Henan and Hebei provinces along the west edge of the Third Terrace, carrying water to Beijing and Tianjin, respectively. With high ground level, the trunk canal flies over other rivers thus avoiding the water of premium quality from being contaminated. The water can flow by gravity through the canal running in parallel and proximity with the Beijing-Guangzhou Railway, supplying water to cities in the Huang-Huai-Hai Plain.

The West Route project is located on the Qinghai-Tibet Plateau, the highest geographical terrace in China. It draws water from the Tongtianhe River, the Daduhe River, and the Yalong River including its tributaries in Upper Yangtze River Basin. These rivers are geographically close to the upper reaches of the Yellow River, and the diversion route goes to the Yellow River crossing the Sichuan and Qinghai provinces. The high altitude West Route has the advantage of a wide coverage of service, mainly supplying water to the upper-middle reaches of the Yellow River and parts of the northwest region, and on occasional and punctual basis supplementing water to the lower reaches of the Yellow River.

The three water transfer routes of the South-to-North Water Transfer Project connects China's four major rivers, namely, the Yangtze River, the Yellow River, the Huaihe River and the Haihe River, to form a macro crisscrossing network of rivers and canals, which as an organic whole are complementary with one another in terms of macro-scale water resources reallocation across geographical regions of the country.

Some local inter-basin water transfer projects have been put forward in the neighboring water-deficient areas that are not included in the coverage of the South-to-North Water Transfer Project, including the Water Diversion Project to Central Yunnan, Hanjiang-to-Weihe Water Diversion Project, and the Yangtze-to-Huaihe Water Diversion Project.

9.2.1 East Route of the South-to-North Water Transfer Project

The East Route of the South-to-North Water Transfer Project is built on the Jiangsu Province's Northward Water Transfer Project by extending it further to North China. It makes use of the existing canals, waterways, lakes and structures of the ancient Grand Canal and those in the Huaihe and Haihe river basins. Its layout takes into consideration multiple purposes including flood control, drainage and navigation. The East Route trunk extends about 1156 km long and its water transfer capacity as planned is 14.8 billion m^3, supplying water to the Jiangsu, Shandong, Anhui, Hebei provinces and Tianjin Municipality for urban domestic, environmental and industrial water uses, with proper care of agricultural and other uses.

The East Route project is planned into three phases of implementation. In the Phase I, the annual average volume of water withdrawal by pumping from the Yangtze River is 8.7 billion m^3, mainly supplying water to the Jiangsu and Shandong provinces. The Phase II project is an expansion of Phase I, extending to Tianjin in the north, with annual average volume of water withdrawal from the Yangtze River being 10.6 billion m^3. The Phase III builds on the previous two phases with enlargement of the river channels on the trunk and the west part of the lateral canal leading to the Shandong Peninsula.

9.2.2 *Middle Route of the South-to-North Water Transfer Project*

The Middle Route project diverts water from the Hanjiang River in the middle Yangtze River Basin to Beijing, Tianjin, Hebei and Henan. The water supply targets urban domestic and industrial water uses, while supplementing agricultural and ecological water use. According to the planning, the project is implemented in two phases, i.e. Phase I and II, with water diversion from the Hanjiang River first and then from the Yangtze River mainstream. In Phase I, water is drawn from the Danjiangkou Reservoir on the Hanjiang River, the largest tributary of the Yangtze River. The canal starts from the Taocha Intake, passes through the dividing of the Yangtze River and Huaihe River basins at the place named the Fangcheng Saddle and enters into the Huaihe River Basin. It crosses the Yellow River in the west of Zhengzhou and heads up to north along the west side of the Beijing-Guangzhou Railway to the destination, the Tuancheng Lake in Beijing. The headrace of main canal to Tianjin is located at the north of the Xiheishan Village in Xushui County of Hebei Province, and the main canal ends up at the Outer Ring River in Tianjin. The total length of the Phase I trunk canal is 1432 km. Water is conveyed through open canals for the section from the intake to the Hebei-Beijing border, while through pipes and culverts further to Beijing and Tianjin. The annual average amount of water transfer is 9.5 billion m^3 for the Phase I project. In the long term, increasing diversions from the Hanjiang River will exceed the carrying capacity of water resources in the Hanjiang River Basin, which needs diversion from the mainstream of the Yangtze River (the Three Gorges Reservoir) to the Hanjiang River. In that case, the annual average volume of water transfer will reach about 13 billion m^3.

9.2.3 *West Route of the South-to-North Water Transfer Project*

The West Route project involves damming on the Tongtianhe River, Yalong River and the upper Daduhe River in the upper Yangtze River Basin. A water conveyance tunnel, which crosses the dividing range of the Yangtze River and the Yellow River, namely the Bayankala Mountains, will be excavated, diverting water from the Yangtze River into the upper reaches of the Yellow River. The target of the West Route project is mainly to address the water shortages in 6 provinces (autonomous regions) including Qinghai, Gansu, Ningxia, Inner Mongolia, Shaanxi, and Shanxi in the upper and middle Yellow River Basin, in which particularly to mention the Guanzhong Plain in the Weihe River Basin. With auxiliary engineering measures, water can also be supplied to the Hexi Corridor in Gansu Province adjacent to the Yellow River, and to the lower reaches of the Yellow River on occasional basis.

The West Route project is planned to be constructed in phases. In the starting phase, waters will be diverted from the Daduhe River, the mainstream Yalong River (at the place of Ada) and its five tributaries (upstream of Ada) to the upper reaches of the Yellow River and its tributary named the Jiaqu River. According to the planning, the annual average amount of water transfer will be 8 billion m^3 in the target year of 2030. In later phase, dams will be built on the Tongtianhe River to divert water to the Yellow River via the Yalong River.

The West Route project lies on the Qinghai-Tibet Plateau, one of the most geologically complicated regions in China, where the earthquake intensity generally ranges 6–7 magnitude, and even up to 8–9 magnitude in some areas. Its implementation involves most sophisticated engineering techniques and high difficulty of construction. According to the Reply from the State Council to the Overall Planning for the South-to-North Water Transfer Project, the preliminary work for its preparation is underway.

9.2.4 Yangtze-to-Huaihe Water Diversion Project

This project is planned to divert water from the Yangtze River mainstream at two locations, namely the Phoenix Neck on the downstream of Hukou and the Zongyang Water Complex, to the middle reaches of the Huaihe River through the Chaohu Lake which spans across the Yangtze River and Huaihe River basins. Its water supply service area involves 46 counties (cities or districts) in 12 cities (Bozhou, Fuyang, Suzhou, Huaibei, Bengbu, Huainan, Chuzhou, Tongling, Hefei, Ma'anshan, Wuhu and Anqing) of Anhui Province and nine counties (cities or districts) in two cities (Zhoukou and Shangqiu) of Henan Province. The project mainly targets at urban and rural water supply and navigation waterway linkage of the Yangtze River and Huaihe River, in conjunction with supplementing water for irrigation and improving the ecological environment in the Chaohu Lake and Huaihe River basins. Once completed, it will effectively end the shortages of water for industrial and domestic water uses in urban and rural areas along the water conveyance line, the riverine area of the Huaihe River and in the Huaibei area; and partially meet water demands for agricultural irrigation as complements. Parts of the water conveyance canal, integrated with navigation channel development, will serve as a second navigation waterway connecting the Yangtze River and Huaihe River basins, a new complement to the network of inland waterways. The water exchange between the Yangtze River and Chaohu Lake improves the water eco-environment of the lake basin, and yields significant comprehensive benefits such as water supply, navigation and ecology.

In 2030, according to the planning, the annual average volume of water transfer by the project will be 3.303 billion m^3. In the long run, in 2040, that number will increase to 4.3 billion m^3.

9.2.5 Yangtze-to-Weihe Water Diversion Project

The project diverts water from the Huangjinxia Reservoir on the upper reaches of the Hanjiang River mainstream and the Sanhekou Reservoir on the middle reaches of the Ziwu River, one of Hanjiang River's tributaries. The water runs into the Weihe River system through the Qinling Mountains. The project targets at supplying water for domestic and productive uses in five key cities (Xi'an, Baoji, Xianyang, Weinan and Yangling in the Guanzhong area), 13 county-level cities such as Xingping and Wugong, and eight important industrial parks along the water transfer line.

The Guanzhong area, the core of economic and social development in Shaanxi Province, has a severe shortage of water resources. It is actually covered as one of the water receiving areas in the planning of the West Route of the South-North Water Transfer Project, which, however, is facing great engineering challenges. As a matter of expediency to meet the urgent water demands in the Guanzhong area, this project is proposed after comparison among many water transfer schemes, featuring such advantages as reliable quantity of water transfer, low engineering difficulty, and high feasibility, believed to be the best solution to alleviating water shortages in the Guanzhong area.

In view of the fact that the Hanjiang River has to meet its own water demands for maintaining ecology and environment after implementation of the Phase I of the Middle Route project, the proposed project of water transfer from the Hanjiang River to the Weihe River is planned to be carried out in stages. In accordance with the planning, the water diversion from the Hanjiang River should be no more than 1.0 billion m^3 in the year 2020. By 2030, water diversion can be increased to 1.5 billion m^3, as in that stage water is diverted from the Yangtze River mainstream to the Hanjiang River as part of the Middle Route later stage of the South-to-North Water Transfer Project; or as an alternative solution, the incremental water transfer to the Guanzhong area can come from the Jialing River.

In order not to substantially affect the Phase I of the Middle Route of the South-to-North Water Transfer, its operation is subjected to the unified dispatching and management of water resources in the Hanjiang River Basin, and measures shall be taken to suspend diversion in low flow years.

9.2.6 Water Diversion to Central Yunnan

The central Yunnan area contains 49 counties (cities or districts) under the respective jurisdiction of Dali, Chuxiong, Kunming, Qujing, Yuxi, Honghe and Lijiang. It is the political, economic and cultural center of Yunnan Province. The central Yunnan area lies in the zone dividing the four major river basins, namely, Jinshajiang River, Lancang River, Honghe River and Nanpan River (draining to the Pearl River). This area features small catchments, low precipitation and high evaporation, and it has been in a grave situation of water shortage. Supplying water to the central Yunnan area is one of the tasks for comprehensive utilization of the upper reaches of the Jinshajiang River. This project draws water from the Shigu Town of the Jinshajiang River. The annual average amount of water transfer is 3403 million m^3. The water supply service covers Dali, Chuxiong, Honghe, Kunming, Yuxi, Lijiang, and so on, meeting water demands for urban domestic and industrial uses, with proper considerations given to agricultural and environmental water uses.

9.3 Practice and Achievements

Facing the grim situation of increasing contradiction between population, resources and environmental protection in water-scarce areas and on the basis of sufficient scientific analytics, the inter-basin water transfers have steadily entered into the engineering practice stage from the blueprint of planning. On December 27, 2002, the first phase construction of the East Route of the South-to-North Water Transfer Project started, and the water started flowing to north on November 15, 2013. On December 30, 2003, the first phase construction of the Middle Route Project started, and the water transfer began on December 12, 2014. On December 29, 2016, the construction of the water transfer project from the Yangtze River to the Huaihe River started, with a total construction period of 72 months. On September 28, 2014, the National Development and Reform Commission officially approved the Feasibility Study Report for Water Transfer from the Hanjiang River to the Weihe River in Shaanxi. On December 20, 2018, the Huangjinxia Hydraulic Complex Project was officially started, with a total construction period of 87 months. On August 4, 2017, the construction of Water Transfer to Central Yunnan Project was officially commenced, and its completion would be 96 months thereafter.

To date, the respective first phases of both the Middle and East routes of the South-to-North Water Transfer Project have been concluded and water has been transferred, exhibiting to China and the world the brilliant achievements of inter-basin water transfer projects which have the widest coverage, the largest volume of water diversion, and the toughest engineering challenges.

9.3.1 *Phase I of the Middle Route of the South-to-North Water Transfer Project*

Since when the Phase I Project was commissioned on December 12, 2014, it has been running smoothly for four years. During this period, safe operation, stable water supply, and above-standard water quality have been achieved, manifesting its premium economic, social and ecological benefits. With the continuous improvement of appurtenant structures in water import areas, the water supply scope has expanded, and the volume of water transfer increased year by year. The number of outlets along the Phase I trunk canal has increased from 33 in the first year to 110 in the fourth year, and the corresponding amount of water diversion at Taocha Intake increased from 2.167 to 7.458 billion m^3. As of February 18, 2019, the Danjiangkou Reservoir had accumulatively contributed more than 20 billion m^3 of water to the water receiving area via the Middle Route for the past four years.

At present, the water transferred by the Middle Route project has become an important source of water supply for many places such as Beijing and Tianjin. Specifically, 39% of Beijing's domestic and industrial water consumption comes from the project, while 73% of the tap water supply in Beijing's main urban districts is sourced from the project. By comparison of the per capita water resources before and after the project, it increases from 100 m^3 per person to 150 m^3. 80% of the total water supply for domestic and industrial water uses in Tianjin comes from the Middle Route project, which covers 14 districts of the Municipality. The previous single source of water supply is water diversion from the Luanhe River, now it is doubled to have water coming from the Middle Route, greatly improving the assurance of water supply. In Henan Province, 37 water-receiving cities and counties have accessed to water supply from the Middle Route. More than 80% of the tap water in the central districts of Zhengzhou attributes to the Middle Route project. The main urban areas of Hebi, Xuchang, Luohe and Pingdingshan are completely dependent on the project for water supply. In Hebei Province, 75% of the water supply in the main urban areas of Shijiazhuang, Handan, Baoding and Hengshui is sourced from it.

The water quality of the Middle Route Project is premium, generally at Grade II or above, which significantly improves the quality of domestic water for the residents in the water-receiving area. The hardness of tap water in Beijing dropped from 380 mg per litre before the water transfer to a range of 120–130 mg per litre after that. More than four million people in Hebei Province bid farewell to the history of using high fluoride water and brackish water.

After the commissioning of Phase I Project, the rapid drop of the groundwater tables in the North China Plain was effectively deterred through a combination of measures including restriction on excessive groundwater pumping, groundwater recharge with the water transferred, and substitution of the misused waters entitled for the environment. The cumulative avoidance of groundwater exploitation was estimated to be about 1.5 billion m^3 and the groundwater table in the plain area has obviously rebounded. By the end of May 2018, the groundwater level in the plain of Beijing had risen by 0.91 m compared with the same period in the previous year.

The groundwater level in 38% of the Tianjin area had rebounded, and 54% of the area had remained stable. The deep formation groundwater in Hebei Province had shifted from annual decrease to an increase of 0.52 m. The groundwater level in the water receiving area in Henan Province had risen by an average of 0.95 m. As the groundwater level rose, the groundwater table funnel became smaller.

At the same time, the Phase I Project of the Middle Route also supplements water resources for ecological uses to several rivers in Henan, Hebei and Tianjin, significantly improving the local ecological environment and landscape. From April to June in 2018, the Phase I Project implemented ecological water replenishment to 30 rivers including Bai River, Qing River, Li River, Fuyang River, Qili River, Hutuo River and Pu River. The cumulative amount of water replenished reached 865 million m^3. The water surface area of rivers, lakes, and wetlands in cities along the route and the water surface of Baiyangdian Lake were significantly expanded, greatly improving the regional water ecological environment and restoring vitality of the water-receiving rivers. Among them, the ecological water replenishment to Baiyangdian Lake totaled 112 million m^3, and the inflow water quality observed at the cross-section was upgraded from the previous below Grade V to Grade II after water is introduced in. The Puhe Reservoir that had been dried up for 36 years is once again filled with water. An increase of water body areas (rivers and canals) in the Xushui District is about 430,000 m^2, and the near-surface groundwater tables in the vicinity of the river channels increase by an average of 0.96 m. This not only improves the water ecological environment on the upper reaches of Baiyangdian Lake, but also improves the water quality and ecological environment of the lake. During the Phase I project, 47 million m^3 of water resources for ecological use was transferred to the important rivers, lakes, and wetlands as well as water-deficient areas in Tianjin. The water ecology of the Haihe River is considerably ameliorated, the groundwater level recovers, and the water quality of the central urban area is improved remarkably. The water quality at the monitored cross-sections of the four rivers is upgraded from Grade III–IV to Grade II–III, and the water quality of other rivers is also improved to varying degrees.

Since the commissioning of the Phase I of the Middle Route project, the water supply structure of areas including Beijing, Tianjin, Hebei, and Henan has been optimized. The water shortage situation has been effectively alleviated, and the degraded ecology and environment resulting from the over-exploitation of water resources have been significantly improved. The project has strongly supported the economic and social development of the water-receiving areas as well as improved the drinking water quality for the people.

9.3.2 Phase I of the East Route of the South-to-North Water Transfer Project

The East Route Project was commissioned on November 15, 2013. During the past 5 years, nearly 3.1 billion m^3 of Yangtze River water has flowed to the north, relieving the water shortage crisis in northern Jiangsu and especially the Jiaodong area for which the water transfer has become an irreplaceable water supply source to secure its urban water uses with preeminent comprehensive benefits.

With the East Route project, the Shandong Province has been given stronger capacity to effectively cope with drought and reduce disasters. In 2014, the four cities in the Jiaodong area suffered a prolonged dry season with long period of low precipitation, and its local water resources were in a crisis. To handle this situation, Shandong Province coordinated projects such as the South-to-North Water Transfer Project, the Water Diversion Project from the Yellow River to the Qingdao City, and the local Jiaodong Area Water Transfer Project, to implement four operations of emergency drought-relief water transfers to the four drought-stricken cities mentioned above. The province also launched water transfers for three consecutive years in the flood season.

The water transfer through the East Route not only secures urban water supply in the four cities of the Jiaodong area, but also alleviates the pressure on the water ecological environment in the water receiving area. On June 21, 2015, the water level of Baotu Spring fell to the lowest level 27.16 m compared to the same period in the years after the spring spewed water out constantly in 2003. From 2015 to 2016, the East Route project, together with other water transfer projects, replenished 58 million m^3 of water in total to Baotu Spring, restoring the scene of active springs in Jinan all the year round.

With the East Route Project in operation, the northern part of Jiangsu Province has eased the pressures on industrial, agricultural, domestic and ecological environment water uses in 50 counties (districts or cities) and 7 cities, with nearly 40 million people being covered. Meanwhile, the project has other functions including flood control, surface drainage, irrigation, and navigation, boosting the socio-economic development in the northern Jiangsu Province.

9.4 Outlook and Development

The inter-basin water transfer project has the following traits: wide coverage, long transfer distance, large scale, long construction period, and complicated water resources allocation and scheduling. In order to give full play to the comprehensive benefits of the project, the planning, construction and operation management of such project shall emphasize optimizing the water resources allocation system, improving the project layout, and enhancing the scientific and technological capabilities for inter-basin water resources management, based on the new situation of water resources change and new demands of economic and social development. In the coming period, we shall focus on promoting inter-basin water transfer from the following aspects.

9.4.1 Improving Project Layout

In the New Era, when we plan and justify the inter-basin water transfer project, we should adhere to the principle of "giving the priority to ecology, promoting well-coordinated environmental conservation and avoiding excessive development", and follow the road of "taking water saving as the priority and pursuing green development". Only through thorough analysis of the spatial equilibrium of water resources and coordinating the water demands in both water import and export areas, can we better protect the ecological environment and determine a sound project scale and implementation scheme.

The General Plan of the South-to-North Water Transfer Project sets the overall pattern of re-allocation of China's water resources, namely, "four latitudinal rivers and three longitudinal canals" referring to the country's crisscrossing water courses for water conveyance. After the completion of the first phases of both Middle and East route projects, remarkable economic and ecological benefits have been witnessed. The Yangtze River, as a strategic water source in the country, takes its due role to better support the national strategies such as the Belt and Road Initiative, the Coordinated Development for the Beijing-Tianjin-Hebei Region, the construction of the Xiong'an New District and the development of city clusters in Central Plain region. In this connection, it is required to advance the preparatory work, including planning and analytics, following-up water transfer from the Yangtze River to the Hanjiang River as part of the Middle Route project, the Phase II of the East Route, and the West Route project, so that construction of them can be commenced in a timely manner, and the overall layout of the South-to-North Water Transfer Project can be fulfilled as envisioned on the earliest possible date.

9.4.1.1 Yangtze-to-Hanjiang Water Diversion Project

With socio-economic development, the central government has laid out a series of nation-level development strategies targeting the North China Plain, such as the coordinated Beijing-Tianjin-Hebei development and Xiong'an New District construction. These strategies have demonstrational significance in promoting innovation of the regional development institutions and mechanisms, patterning urban agglomerations, charting the pathway to building ecological civilization, and boosting the coordination among population, economy, resources and environment. The plan that derived from the aforesaid strategies clearly requests breakthroughs to occur as early as possible and specifies the Year 2035 as deadline to come at all the strategic planning targets, for which water resources are a supporting pillar. According to the projection, the gap of water supply will be up to more than 4 billion m^3. Therefore, it is urgent to fill the gap of water resources.

In the fourth year after the Phase I Middle Route project was put into service (2017–2018), the volume of water intake at Taocha has reached 7.458 billion m^3, which is 78.5% of the annual average design capacity. During 2018–2019, the registered amount of water transfer to the water-receiving area is 9.452 billion m^3 (at Taocha cross-section), basically equal to the designed annual average diversion, but it happened much earlier than the planned date of reaching full diversion capacity, which is ten years after the commissioning according to planning. Affected by the variability of inflows into the Danjiangkou Reservoir, if the Middle Route Project increases diversion from the Hanjiang River to support the Beijing-Tianjin-Hebei region, the Xiong'an New District and other national development strategies, there will be such problems of unreliable water sources and low assurance rate. To address this situation, according to the General Plan of the South-to-North Water Transfer Project, Yangtze-to-Hanjiang Water Diversion Project, as supplements of the water source, shall be commenced in a timely fashion.

According to the plan, once the Yangtze-to-Hanjiang Water Diversion Project is completed, the water transfer through the Middle Route can be increased by about 1.4 billion m^3 annually under the condition of not enlarging the trunk canal. This also enhances the capacity of the Hanjiang River to supply water to the Weihe River as well as to downstream reaches of the Hanjiang River itself with additional releases. As such, the existing trunk canal of the Middle Route Phase I Project can operate to its full capacity much earlier than expected, and thus maximizing its effectiveness and benefits.

In the longer term, if the existing trunk canal is enlarged locally or in full length, the annual water transfer to north can see an increase by 2.8–3.5 billion m^3.

9.4.1.2 Phase II of the East Route of the South-to-North Water Transfer Project

With the rapid development of China's socio-economy, the water demands and engineering conditions in the service area of the East Route project as specified in the General Plan of the South-to-North Water Transfer Project have undergone great changes. New development concepts and water governance ideas impose higher requirements on the East Route project. Especially after the implementation of the coordinated Beijing-Tianjin-Hebei development strategy, the region articulated new demands for the East Route project. In 2012, the sixth meeting of the State Council's South-to-North Water Transfer Project Construction Commission explicitly requested that the planning analytics for the follow-up projects for the East and Middle Routes be accelerated. Therefore, it is advisable to consider the necessity and possibility of merging the implementation of the second and third phases of the East Route project, according to the actual conditions and new requirements of water supply areas. Further analytics and justification for the Phase II and III project planning are required, in a bid to scientifically determine the project scale and optimize the overall layout.

9.4.1.3 West Route of the South-to-North Water Transfer Project

On the basis of the General Plan for the South-to-North Water Transfer Project, the Ministry of Water Resources has carried out thematic researches on issues related to the West Route project since 2012. The researches respond to the deployment made at the sixth plenary meeting of the State Council's South-to-North Water Transfer Project Construction Commission, take the new situation and tasks into account, and focus on the necessity of this project and the impacts of water diversion and transfer. In September 2016, the National Development and Reform Commission submitted to the State Council the Report by the National Development and Reform Commission on the Relevant Issues Concerning the West Route of the South-to-North Water Transfer Project. The report put forward the ensuing research and analytical programs which shall be comprehensive, profound and systematic, based on the opinions or suggestions proposed by relevant provinces, ministries, and experts on major issues,.

In the context that the CPC Central Committee comes up with the new development concepts quoted as "innovative, coordinated, green, open, sharing" as well as the guidelines of water governance, namely, "prioritize water conservation, balance development with spatial distribution of water resources, practice systematic governance and achieve government-market synergy", and given the changing situation of inflow and sediments of the Yellow River and the basin's socio-economic status, it is suggested that 3 aspects of work need to be strengthened as follow-ups: First, deepening the researches and analytics on major and key issues concerning the West Route project; second, strengthening comparative studies and analytics on the alternatives of the West Route project; third, reinforcing water resources management and water ecological restoration in the Yellow River Basin.

9.4.2 Strengthening Integrated Management

Several river basins and numerous water-receiving areas are usually involved in the inter-basin water transfer. It is a complex and systematic project featuring large scale, long conveyance distance, large water supply coverage, and complicated construction and operation management. Only by means of enhancing integrated planning and management of water resources in the stage of implementation, can the sustainability of water resource utilization be secured and the expected comprehensive benefits for the society of such project be fulfilled. Therefore, it is vital to develop the cross-basin integrated and unified water resources dispatching platform with pertinent institutions and mechanisms formulated and operational.

9.4.2.1 Institutions for Rational Inter-basin Allocation of Water Resources

The preparation, construction and operation of the inter-basin water transfer project involve various groups of interests. Whether the concerns and interrelations of all these parties can be properly addressed is decisive to the successful implementation and smooth operation of such project, especially the interrelation between the central and local governments as well as between the water export and import areas must be well harmonized. Following holistic coordination approaches, we shall strengthen the unified planning and macroscopic readjustment of water resources allocation, customize and update water allocation plans in a timely manner contingent on the dynamics of water supply and demand as well as national economic and social development plans.

9.4.2.2 Mechanisms of Ecological Water Use and Compensation for Ecological Service

The water ecological environment will be subjected to changing after water is transferred out of the source basin. Therefore, the water diversion in the planning should be within the acceptable range of the local ecological environment to ensure that the ecological water demands in the source basin itself are met. Monitoring on the ecosystems being potentially impacted by the transfers as planned shall be strengthened, with compensation and remedial measures taken to mitigate the impacts. The goal is to foster co-development for both water import and export areas.

9.4.2.3 Mechanisms Establishment of Paid Use and Economic Adjustment for Inter-basin Water Transfer

The water resource valuation system needs to be developed, which shall help facilitate the rational development, utilization, conservation and protection of water resources, promote the coordination among water resources utilization, socio-economic development as well as eco-environmental protection, and fulfill the due comprehensive benefits of water resources. The water pricing mechanism shall conform to the principles of cost compensation, reasonable charging of fees, better quality with higher price, and fairness in charging, which as an economical instrument can prevent such phenomena of squandering water resources and the resultant more pollution.

9.4.3 Researches Enhancement

Endowed with relatively rich water resources in the Yangtze River Basin, the water transfer from the Yangtze River to other basins has great strategic significance for the north–south re-distribution of China's water resources. Generally speaking, the volume of water transfer in the planning does not have notable negative impacts on water uses in the Yangtze River Basin, justifying the feasibility of such water transfer projects. However, in dry years or seasons, the superimposed impacts brought about by the simultaneous operation of all the water transfer projects can no longer be ignored. In addition, the safe and efficient operation of the water transfer projects themselves is of a big concern as well, especially the long-distance water conveyance system, for which the key technologies of control and operation need to be further studied. From the long-term perspective, only the win–win outcomes for both the water export and import areas can ensure the long-term effectiveness and benefits of such water transfer project, therefore, profound researches on measures to counter the eco-environmental impacts on the water source area and on ecological compensation mechanism are much needed.

9.4.3.1 Study on the Superposition Effect of Inter-basin Water Transfer Projects

As far as a single inter-basin water transfer project is concerned, it has minimal impact on the water resources utilization and ecological environment in the Yangtze River Basin. However, with many such water transfer projects working to divert from the Yangtze River Basin, their superimposed impacts are not negligible, especially in dry years or seasons, such as the intrusion of tidal sea water in the Yangtze River estuary and the impact on downstream water uses and on the ecological environment. Therefore, it is a matter of urgency to launch the researches on the cumulative and superimposed impacts of all the inter-basin water transfer projects in operation, emphatically studying the joint water supply dispatching plan integrating the Three Gorges Reservoir as the core, key Hanjiang River reservoirs and all the inter-basin water transfer projects. This type of joint operation will improve the capacity of the Yangtze River Basin in re-allocating water resources, emergency response, and converting floods into water resources.

9.4.3.2 Key Technology Research on Safe and Efficient Control and Operation of Water Transfer Projects

The operation and control of the water transfer project involves multi-disciplines and cutting-edge inter-disciplinary theories, including hydraulics, automatic control theory, hydraulic hammer effect prevention and protection theory, and computational science. In China, the large-scale water conveyance and distribution project, with the Middle Route of the South-to-North Water Transfer Project as an example, has its automatic control system developed. However, the system is only specifically applicable for controlling the large cross-section, low velocity, open-channel and gate-regulated flows. Many other water transfer projects, such as the Water Diversion Project to Central Yunnan, Hanjiang-to-Weihe Water Diversion Project, and Yangtze-to-Huaihe Water Diversion Project, are under implementation in tight schedules. These projects involve the operations of pumping stations and large-scale pressure conduits, incidents of alternating non-pressure and pressure flows or pressure operation in free flow tunnels, as well as the disturbance of diversion intake gates under autonomous operation condition. All above mentioned have a great impact on the safe operation of the projects, and thematic research needs to be carried out.

9.4.3.3 Research on Ecological Environment Impact Measures and Ecological Compensation Mechanism

In order to sustain the long-term operation and benefits of the water transfer project, one must properly manage the inter-relations between the water import and export areas, and among resources, environment and socio-economic development. Although thematic studies on the impacts of water diversions have been conducted for each of such projects, the impact on the ecological environment manifests itself over a long-term period and a rather complex process. Therefore, even in the operation stage, the mitigation measures, risk control and ecological compensation mechanism also need to be studied and implemented, to ensure the sustainable availability of water resources in the water source basin.

Chapter 10
Joint Operation of the Group Reservoirs

To mitigate the flood and drought disasters, develop and utilize the water resources, and protect the water eco-environment, a number of key water resources and hydropower projects have been planned on the main stem and major tributaries of the Yangtze River, most of which are distributed on the Jinsha River, the Yibin-Yichang stretch of the Yangtze River mainstream, the Yalongjiang River, the Minjiang River, the Daduhe River, the Jialingjiang River, the Wujiang River, the Qingjiang river, the Hanjiang River, the Dongting Lake tributaries, and the Poyang Lake tributaries, etc. These projects have multiple objectives including flood control, power generation, navigation, and water supply. Going along with China's socio-economic development and step-wise implementation of the Master Plan of the Yangtze River Basin, a group of reservoirs with the Three Gorges Reservoir as the core have been constructed up to date. According to statistics, more than 300 large-scale reservoirs, each having a storage capacity over 100 million m^3, have been built in the Basin, with a total regulation capacity of more than 180 billion m^3 and an aggregate flood control capacity of some 77 billion m^3.

10.1 Significance

The key water resources projects in the Yangtze River Basin are huge in magnitude and far-reaching in terms of their influences. The operations of these huge projects have to take account of such highly-correlated multiple objectives as flood control, water supply, power generation, navigation, water-related ecosystem and environment, and so on, which are either agreeable or contradicting with one another in the enormous and complex system. Conducting joint operation of the group reservoirs can further ensure basin-wide security of flood control, achieve higher efficiency in hydropower generation and comprehensive utilization of water, safeguard the domestic, productive, and ecological water uses, improve navigability of the Yangtze

© Changjiang Press (Wuhan) Co., Ltd. 2022

Changjiang Water Resources Commission of Ministry of Water Resources of the People's Republic of China, *Report on Yangtze River Rehabilitation and Protection 2019*,
https://doi.org/10.1007/978-981-16-4927-1_10

River waterways, maintain healthy river and ecosystem conditions, and secure the development of the Yangtze Economic Belt and operation of the Golden Waterway, thereby promoting the sustainable and coordinated development across the entire Basin.

10.1.1 Joint Operation of Group Reservoirs in the Yangtze River Basin Can Further Secure Safety of Flood Control in the Basin and Relieve Flood Pressure on the Middle and Lower Reaches

The upper reaches of the Yangtze River are the main source of the Yangtze River floodwater. The collection of flood discharges from the upper mainstream namely the Chuangjiang Reaches and from tributaries within the territories of the Sichuan Province and Chongqing Municipality, makes its way downstream, passing Yichang to the middle and lower reaches, and transmitting along with the rushing floods is the formidable pressure for flood control. The large cascade reservoirs on the upper Yangtze River mainstream and tributaries, with the Three Gorges Reservoir serving as the backbone, are important components of the Yangtze River flood control system. The joint operation of these reservoirs can effectively improve the overall flood control capacity in the basin as a whole by taking advantage of the differences in regulation performance of individual reservoirs to reduce and stagger flood crests in a well-coordinated manner, which work in synergy to regulate and store floodwaters. The benefits with respect to flood protection are several folds: to secure the safety of the reservoirs themselves and the safety of flood control in the upper basin, and to lessen the pressure of flood control on the middle and lower reaches.

10.1.2 Joint Operation of Group Reservoirs in the Yangtze River Basin Can Further Secure Water Resources Supply

By means of making full use of the storage capacity compensation among individual reservoirs, the joint operation of them is able to improve water supply capacity in dry season, thus safeguarding the fundamental security for domestic, industrial, agricultural and ecological water uses. Particularly in the scenario of unusually low runoff in the Yangtze River as a consequence of climate change with onset of local-scale drought disasters, the joint operation can improve water supply security during unusually dry years or dry seasons. In extremely dry years, the joint operation of the group reservoirs is one of the most important disaster mitigation measures to meet the emergency demand for water supply.

10.1.3 Joint Operation of Group Reservoirs in the Yangtze River Basin Can Further Optimize the Allocation of Water Energy and Promote Development of Green Energy Resources

As a significant hydropower generation base to feed the West-to-East Power Transmission, the group of hydropower stations in the Yangtze River Basin, with their great power supply capacity and the trans-regional and trans-grid characteristics, can realize the allocation and distribution of electricity over a larger scope of regions. By joint operation, the hydropower stations co-ordinate with each other to better regulate the streamflow through making full use of their respective storage capacity. Without increase of the running cost of those hydropower stations, the joint operation can simultaneously enhance the stability and adaptability of both the water regulating and power regulating systems. Especially in the years of less-than-average runoff or unusually disproportionate inflows coming from the Yangtze River mainstream and tributaries, the joint operation proves to raise the level of power supply security.

10.1.4 Joint Operation of the Group Reservoirs in the Yangtze River Basin Can Further Maintain River Health and Mitigate the Impacts of the Engineering Works

The construction and operation of the reservoirs change the natural hydrograph and hydraulics of the river, which may have consequences to a certain degree on the intact eco-environment of the river. The previous national code on reservoir operations in China emphasized the economic benefits such as flood control, power generation and water supply, as well as societal safety, but took insufficient care of the ecological and environmental demands of the river itself. In this regard, damming the river may have imposed certain impacts on the water-related ecosystem. Practical experience shows that the joint operation of the group reservoirs can help create suitable hydrological and hydraulic conditions for important aquatic animals in the river to spawn, breed and grow by means of simulating natural hydrological regime, thereby reducing adverse impacts of engineering construction on river ecology through compensating for or even restoring the river ecology downstream.

10.2 Evolution Process and Effectiveness

Given the huge scale of the group reservoirs in the Yangtze River Basin, the goal of the joint operation of these works in the New Era is to formulate a science-based scheme of joint operation and conduct such joint operation on real-time optimization basis, so as to better synergize functioning of all the component reservoirs in the group for concerted regulation of floods, proper allocation of the limited water resources over time and space to meet water demands for various uses including the river ecosystem itself. In line with the deployment by the Ministry of Water Resources of China, the Changjiang Water Resources Commission has carried out a series of research studies on the joint operation of the group of key reservoirs on the Yangtze River mainstream and tributaries with the Three Gorges Reservoir as the focal point. Meanwhile, the CWRC has also collaborated with the operation and management corporations of the water projects, e.g. the Three Gorges Project Corporation (Group), to study on the dispatching of particularly important reservoirs and river reaches. With these research findings, the CWRC, since 2012, has organized the development of annual scheme for joint operation of the group reservoirs located in the upper and middle Yangtze River Basin. The operation schemes that the CWRC has developed and/or amended with approvals from higher levels of authorities include the Scheme for the Yangtze Flood Control approved by the State Council, and the Scheme for Flood and Water Dispatching of the Hanjiang River approved by the State Flood Control and Drought Relief Headquarters. Besides, the CWRC has issued flood control schedules for a number of local rivers, including the Scheme for the Chuhe River Flood Control, the Scheme for the Shuiyangjiang River Flood Control, and the Scheme for the Wujiang River Flood Control. Years of joint operation practices demonstrate that the scientific researches constitute the very foundation on which to implement joint operation and are the prerequisites for fine-tuning the joint operation schemes.

In principle, all the important large-size reservoirs in the Basin should be incorporated into the system of joint operation for floods and water quantity regulations. However, considering such factors as the reservoirs' scale, flood control capacity, regulation capacity, catchment area that the reservoir controls, and its operating condition, the actual scale and number of reservoirs incorporated in the joint operation system keep varying every year. The following stages are identified to describe the evolution process that the joint operation of the group reservoirs has undergone.

10.2.1 Process of Evolution

10.2.1.1 Years from 2012 to 2016

Based on plenty of scientific researches, for the first time ever in history, the joint operation of 10 important reservoirs (hydropower stations) was implemented in 2012, integrating the dispatching of the reservoirs namely the Three Gorges Reservoir together with the Jin'anqiao, Ertan, Zipingpu, Pubugou, Goupitan, Silin, Pengshui, Bikou and Baozhusi reservoirs. In doing so, the principles and objectives of joint operation of this group of reservoirs were set, the joint operation scheme formulated, and the competences of involved institutions clearly defined. The 2012 Scheme for Joint Operation of the Upper Yangtze River Group Reservoirs, the very first scheme of its kind, was approved by the State Flood Control and Drought Relief Headquarters in June 2012, setting a milestone for the Yangtze River Flood Control and Drought Relief Headquarters and the CWRC to carry out joint operation of the upper Yangtze River reservoirs. Thereafter, following the dynamics of reservoir development and their operational status, the number of reservoirs contained in the group expanded to 17 in 2013, and further to 21 key reservoirs in the upper Basin during 2014 to 2016 namely: the Liyuan, Ahai, Jinin Jini, Longkaikou, Ludila, Guanyinyan, Xiluodu, Xiangjiaba on the Jinshajiang River; the Jinping I and Ertan on the Yalongjiang River; the Zipingpu and Pubugou on the Minjiang River, the Bikou, Baozhusi, Tingzikou, and Caojie on the Jialingjiang River; the Goupitan, Silin, Shatuo and Pengshui on the Wujiang River; and the Three Gorges Project on the mainstream Yangtze River. The items being regulated under the joint operation schemes have become increasingly comprehensive year by year, and the annual joint operation scheme as approved has served as a good guideline directing actual operating practice in that it helped maximize such comprehensive benefits of the group reservoirs as flood control, power generation, water supply, navigation and ecology.

10.2.1.2 Years from 2017 to 2018

In 2016, a major flood event occurred in the Middle-Lower Yangtze River. In the actual process of flood management, it was found that the reservoirs on the Qingjiang River and on the tributaries of the Lake Dongting functioned efficiently enough in intercepting the flood flows and staggering flood peaks to lessen the pressure on the middle and lower reaches. Therefore, in 2017, the scope of joint operation of the group reservoirs were accordingly extended downward to the location of the Chenglingji Station, adding in 7 reservoirs on the Qingjiang River and the tributaries of the Lake Dongting in the middle Yangtze River Basin, with the total number of reservoirs coming to 28. In 2017 when the floods in the Middle-Yangtze River hit, the group of reservoirs in joint operation worked so well in flood control and disaster reduction. In 2018, the group reservoirs further expanded to take on board another 12 reservoirs including the Danjiangkou on the Hanjiang River and the Zhelin on the tributary of Lake Poyang, with the number of reservoirs in the group totaling 40, with a total regulation storage capacity of 85.4 billion m^3 and a total flood control storage capacity of 58 billion m^3.

10.2.1.3 Year 2019

With progressive maturing of the integrated system for the Yangtze River flood control and drought relief, the number and scale of drainage pumping stations and flood sluices in the middle and lower Basin are booming and their impacts on the streamflow of the Yangtze River mainstream become no longer negligible. The CWRC plans to eventually incorporate, in the joint operation system, all types of water works, including major reservoirs, diversion works, major pumping stations and sluices, and flood storage and detention basins in the Yangtze River Basin. On the basis of the 2018 Scheme of Joint Operation of the Group Reservoirs and with reference to relevant study outcomes, the CWRC is organizing the preparation of the 2019 Scheme of Joint Operation of Key Water Resources and Hydropower Projects, to be submitted for approval following stipulated procedures. The Scheme being developed takes in, in addition to those reservoirs, 46 flood storage and detention basins, 10 major pumping stations and sluices equipped with flood reporting facilities, and 4 diversion intakes.

10.2.2 Work Done

Starting from 2008, when the Three Gorges Project and the Danjiangkou Reservoir only were operated in synergy in the Yangtze River Basin, it has been contrived to achieve a joint operation mode for group reservoirs. For a decade, great effort have been paid on the aspects of institutional mechanism setup, scale and scope determination, system development, and capacity building. Until today, a well-defined and effective mechanism featuring co-development, technical synergy, high efficiency and smooth coordination among the involved institutions has been in place and functioning.

10.2.2.1 Basic Researches

Since 2009, based on a large number of previous studies on the dispatching of individual reservoirs, the CWRC has carried out basic researches on integrated operation of a group of key reservoirs on the mainstream and tributaries of the Yangtze River with the Three Gorges Reservoir as the core. The Top-level Design for Researches on Joint Operation of Water Projects in the Yangtze River Basin was prepared, and then revised in 2019 to guide the research studies on joint operation over a period of time in the future. In addition, under the framework of closer collaboration between administration and enterprises on win–win basis, the CWRC, in cooperation with the Three Gorges Corporation (Group), has carried out the Phase I (4 years, 27 research items) and Phase II (4 years, 49 research items under 11 subjects) research studies on key technology for scientific operation of the Three Gorges Reservoir since 2011 and is continuing to plan the Phase III studies at present. In collaboration with power generation companies such as the Wujiang Corporation and State Power Corporation, the CWRC has conducted studies on flood scheduling on main tributaries including the Wujiang River, the Daduhe River and the Yalongjiang River. The outcome of the research on Key Technology in Scheduling the Group Reservoirs in the Yangtze River Basin for Integrated Operation of Flood Control and Benefit Promotion, in which the CWRC took the lead, won the Special Award for Scientific and Technological Advance of Hubei Province (2017).

10.2.2.2 Joint Operation Scheme

On the basis of the study outcomes, the CWRC, since 2012, has organized formulation of the annual scheme for joint operation of the group reservoirs in the Yangtze River Basin, containing the principles and objectives of the joint operation, the operation schemes for each individual reservoir in the group, and the competences of involved institutions in running these reservoirs. Such annual schemes, as they evolved over the process, have progressively zoomed in the provisions of the operating procedures and emergency response, handling flood scheduling during wet season, impounding toward the end of wet season, releasing water to supplement downstream water uses during dry season, and drawing down reservoir before floods come. From 2012, the CWRC has developed the flood control schemes and schedules, with approvals from different levels of authorities, such as the Scheme for the Yangtze Flood Control approved by the State Council, and the Scheme for Flood and Water Dispatching of the Hanjiang River approved by the State Flood Control and Drought Relief Headquarters. Besides, the CWRC has also issued flood control schedules for other particular tributaries of the Yangtze River, including the Schedule for the Chuhe River Flood Control, the Schedule for the Shuiyang River Flood Control, and the Schedule for the Wujiang River Flood Control. To date, the scheme for joint operation of the group reservoirs in the Yangtze River Basin has evolved to be more and more matured and operational to guide the actual practice of reservoir operations and management with remarkable social, ecological and economic benefits attained.

10.2.2.3 Capacity Building

The Yangtze River Flood Forecasting System has been developed to cover the whole Basin. The system incorporates more than 13,000 hydrological and rainfall stations, 341 forecast stations, 695 sets of forecast scenarios, 342 gate-regulated reservoirs and more than 100 sets of scheduling scenarios, and have such functionaries as real-time monitoring, flood control situation analytics, hydro-met information query, optimization of operation based on forecast inflow, interactive analysis, reporting and display for presentation, user customization, as well as other hydro information products, capable of integrating the forecasting, scheduling and conferencing consultation processes. Over the years it has played a vital role in flood scheduling operations. The Overall Implementation Plan of the Integrated Operation System for Key Water Projects in the Yangtze River Basin and the feasibility study report for the basic information platform of this system were approved by the Ministry of Water Resources. The project development leading group and the executive office were set up in CWRC and the work plan for the project development drawn up. From 2014 on, the CWRC has organized the management units of key reservoirs in the upper and middle Basin to develop an information sharing platform for joint operation of the group reservoirs, through which the 40 reservoirs incorporated are sharing information.

10.2.2.4 Management Institutions and Mechanisms

In accordance with relevant laws, regulations, and departmental responsibilities, and with the Ministry of Water Resources taking the leadership, the CWRC, building on the existing flood commanding architecture, has established competence-clear and highly efficient institutional arrangement for joint operation of the group reservoirs. This institutional arrangement is featured by "governments take lead, departments co-operate, and enterprises participate" are enabling trans-regional, interdepartmental, and cross-sectoral consultation and collaboration, and promoting the unified scheduling and commanding of water projects in the Basin. With approval by the Ministry of Water Resources in 2016, the CWRC set up the Bureau of Reservoir Joint Operation in conjunction with the existing Flood Control and Drought Relief Office (now renamed as the Bureau of Flood and Drought Disaster Prevention), responsible for the joint operation of key reservoirs in the Basin. Annual symposium on joint operation of the group reservoirs is held, which serves as a communication platform for both administrators and corporations concerned. To get the joint operation of reservoirs on the legal bases, studies have been launched to facilitate the formulation of the Management Rules for Joint Operation of the Group of Key Reservoirs in the Yangtze River Basin and the specific management rules for operation of individual reservoirs including the Danjiangkou, Xiluodu and Xiangjiaba.

10.2.3 Effectiveness

Under the leadership of the Ministry of Water Resources, the CWRC, through strengthened monitoring and forecasting, intensified expert consultation and scrutiny, and joint operation of key reservoirs inclusive of the Three Gorges Project, has provided strong safeguards for the security of flood control, ecology and water supply over the basin and obtained the due comprehensive benefits out of group reservoirs.

In terms of flood control, the largest flood event ever since the Three Gorges Reservoir was built, with peak inflow of 71,200 m^3/s, was safely handled in 2012 through joint operations. In the process, at the most critical moment the Three Gorges Reservoir reduced the flood discharge by as high as 26,200 m^3/s, and cumulatively retained 20.05 billion m^3 floodwater in the reservoir from releasing downstream, lowering the water level at Shashi and Chenglingji by 1.5–2.0 m as compared to the case without such regulation. In 2016, the region-scale floods in the middle and lower Basin were successfully coped with, during which the group of reservoirs retained totally 22.7 billion m^3 of floodwater, reducing the peak water level around Chenglingji by about 1.0 m, thence preventing more than 33,000 hm^2 of farmland from inundation and avoiding evacuation of more than 380,000 people. In 2017, the floods on the middle reaches was effectively controlled, with the group reservoirs working together to store about 14.4 billion m^3 of floodwater, reducing the peak flood levels in both the Dongting Lake and on the mainstream (Lianhuatang section) by about 1.0–1.5 m and shortening the higher-than-critical level period in the Dongting

Lake by about 6 days. In 2018, large floods on the upper reaches of the Yangtze River were effectively dealt with, when a total of 11.1 billion m^3 of floods were detained by the group reservoirs, effectively preventing water levels on the vulnerable reaches of Jialingjiang River from rising up to the critical level and the lower reaches of Minjiang River and the Jingjiang stretch of the mainstream Yangtze from exceeding warning levels. As an important part of the flood control system of the Yangtze River, the group reservoirs in the Basin have played a significant role in storing floods and reducing flood peaks, which have effectively alleviate the pressure of flood control on the upper reaches of the Yangtze River, the Dongting Lake and the middle and lower reaches of the Yangtze River, and as a result improved the overall flood control capacity of the Yangtze River Basin.

In terms of water supply, by enhancing real-time dispatching, the CWRC optimizes allocation of the reservoirs to allow flood control compensating regulation for the middle and lower Yangtze, controls the water levels on dynamic basis, implements flood detention operations for the medium and small floods to promote effective utilization of the flood water as resources. The group reservoirs impound water at the proper timing and in an orderly sequence and planned manner, so as to mitigate the impact of impounding on the downstream. Since 2012, the upstream reservoirs have essentially reached the their respective annual water impounding targets, except for some year-to-year storage reservoirs, particularly in that the Three Gorges Reservoir has, in the trial impounding stage, been impounded to the NPL of El.175m for 9 consecutive years. After 2009, the upper Yangtze River reservoirs have cumulatively supplemented over 200 billion m^3 of water to the middle and lower reaches during dry season, which effectively enlarged the low streamflow and secured water supply downstream. For instance, during the dry season requiring the group reservoirs to supply water in 2017 to 2018, the Three Gorges Reservoir increased water releases by a total amount of 23 billion m^3 to downstream in duration of 177 days.

In terms of ecology, the Three Gorges Reservoir has performed 12 eco-friendly operation experiments in 8 consecutive years starting from 2011, which have facilitated the spawning and reproduction of the fish species that lay drifting eggs. According to the monitoring data, a total of 712 million fish spawns were laid during the experimental operations, accounting for about 40% of the total spawns produced in the 7 years. After 2013, the sediment regulating operations has been undertaken for the Three Gorges Reservoir, with no cumulatively incremental sedimentation observed at the tail of the Reservoir. Starting from 2017, the eco-friendly operation experiments on selective water withdrawal have been conducted for the Xiluodu Reservoir, which is the first attempt that has ever been taken to mitigate the temperature hysteresis effect of water works in the Basin. Meanwhile, the eco-friendly joint operation of the group reservoirs including the Xiluodu, Xiangjiaba and Three Gorges reservoirs has been implemented. The Danjiangkou Reservoir places the ecological benefit high in the list, and has operated as emergency response for many a time to curb the algal blooming on middle and lower reaches of the Hanjiang River. The first ever experimental eco-friendly joint operation, conducted in 2018, of the Danjiangkou Reservoir together with downstream cascades, demonstrated very good ecological effect.

In terms of power generation, the flood water has been made better use of as resources under the precondition that the flood control security is ensured. The group reservoirs in the upper Yangtze River Basin have increased electricity generation by about 60 billion kW, equivalent to saving 21.6 million tons standard coal or reducing 56.9 million tons greenhouse gas emission since the joint operation of the group reservoirs in the Yangtze River Basin commenced in 2012. The yearly power output in 2014 by the Three Gorges Project reached up to 98.8 billion kW, setting a new world record of annual power output by a single hydro-power station. In 2017, the Three Gorges Project alone increased power output by 10 billion kW due to the joint operation of the group reservoirs. In 2018, the power generation of the Three Gorges Project reached 101.6 billion kW, hitting another new high.

In terms of navigation, the joint operation sharply cuts the peak flood discharge in flood season, thus greatly improving the safety of ships. In dry season the joint operation increases the water depth by 0.5–1 m in the downstream channel, and improves 660 km navigation channel on the Chuanjiang section of the upper mainstream, greatly improving the Yangtze River navigation condition. Every year in flood season, the vessels if jammed around the shiplock of the Three Gorges Project are facilitated to pass through by means of real-time optimization of reservoir operation. In dry season, the Three Gorges Reservoir is, at times of necessity, operated to enlarge water releases for alleviating downstream navigability stress. Each year from 2014 to 2018, the annual freight of goods going up and down through the shiplock of Three Gorges Project exceeded 100 million tons, and particularly the annual freight passage in 2018 reached the unprecedented 142 million tons. All in all, the joint operation has effectively promoted the waterway navigation and coordinated economic growth along the river.

In terms of emergency response, the emergency operation for water quantity is conducted in parallel with the joint operation for flood control. In 2015, after coordinating with the State Grid for emergent modification of the power generation plans of relevant hydro-power stations, the CWRC ordered to reduce releases from the Three Gorges Reservoir to create favorable conditions for rescuing the sunken Oriental Star Liner. In the beginning of 2018 and of 2019, under the situation of prolonged rainy, snowy and low air temperature days as well as sea water intrusion at the estuary, the CWRC ordered to increase the daily average release from the Three Gorges Reservoir to 10,000 m^3/s, which effectively ensured the power supply and the security of the power grids, and ease the water supply stress in the estuary region. In 2018, toward the end of the flood season, two events of dammed lakes took place one after the other within one month at Baige on the Jingshajiang River. On the basis of scientific analysis and experts' consultation, the CWRC issued 8 operation orders to reserve the room of the reservoirs, say 1.3 billion m^3 out of the reservoirs on the middle reaches of the Jingshajiang River, to accommodate the potential floods coming down once the barrier might break.

10.3 Prospect on Joint Operation of the Group Reservoirs

Despite of the joint operation system of group reservoirs in Yangtze River Basin being put into use, there is still a gap between the capability of such system in its inception phase and the actual demands of the Yangtze River Economic Zone, the socio-economic development and the law-based administration of the river basin authority. Scientific researches on the joint operation of the group reservoirs in the Yangtze River Basin are still in the primary stage and the understanding on the potential impacts of the reservoirs needs to be deepened. Numerous technical difficulties and management issues wait to be solved in order to promote the comprehensive benefits of multiple objectives including flood control, water supply, power generation, navigation and ecology over the entire basin.

The joint operation of the key reservoirs shall integrate all the requirements for basin-level flood control and drought relief, water resources utilization, as well as water resources and ecosystem protection. It shall target at the comprehensive benefits of the group reservoirs as a whole and mitigate the negative impacts of the less coordinated operation of the key reservoirs. In line with priority order that flood control over other beneficial uses, regulation for water quantity over regulation for power, comprehensive interests over sectoral interests, basin-wide considerations over local considerations, such joint operation system of the group of key reservoirs should be made fully fledged with broad-based participation of all stakeholders through improving laws and regulations, institutional mechanism, technical support and supervisory management.

10.3.1 Improving Laws and Regulations

Currently the national laws and administrative regulations relative to the joint operation management of Yangtze River reservoirs include Water Law, Flood Control Law, Water Pollution Prevention Law, Flood Control Regulation, Drought Relief Regulation, etc. The joint operation and management of key reservoirs involve multiple departments and cross-jurisdictional stakeholders, on which, however, there has yet to be full-coverage, systematic and clear provisions in the currently applicable laws and regulations. Therefore, it is imperative to formulate sectoral decrees and administrative regulations in compliance with the Water Law and the Flood Control Law, especially to enact the Rules on the Management and Operation of Key Water Engineering Works in the Yangtze River Basin to specify the scope, purposes, implementing bodies, supervision and management of operations and the division of responsibilities and competences. Against the backdrop of the ongoing legislating process of the Yangtze River Protection Law which is high on the agenda of the 13th National People's Congress, the following aspects, after being adapted, are meant to be enacted in the Yangtze River Protection Law as articles and provisions. These include the scope, purpose, implementing body, competence, and procedure of the reservoirs' joint operation for flood control, beneficial uses, ecology and emergency as well as the mechanisms of information sharing, interest compensation, risk control and supervisory management.

10.3.2 Enhancing Management Institutions

The management of joint operation of the group reservoirs relates to not only the responsibilities, rights and obligations of many State departments, local governments and reservoir management units but also the social, eco-environmental and economic benefits that the reservoirs may generate. The management institutions for joint operation of the key reservoirs in the Yangtze River Basin can only grow matured through building up empowered and efficient policy-making authority, competent managerial agency, and competence-clear implementing body.

The Ministry of Water Resources is responsible for the flood control and drought relief operation and water quantity emergency regulation as to the major rivers and lakes as well as the key water engineering works. On the basis of the current institutional framework and fitted into the new round of institutional reform, the CWRC, under the leadership of the Ministry of Water Resources, supposes to perform the mandate of joint operation of the basin-wide key water works with the participation of the provincial water resources departments, the line departments for power and navigation sectors and the water works operation and management units.

10.3.3 Establishing Management Mechanism

The management of joint operation of group reservoirs requires technical coordination, interests balancing, information sharing and public participation.

10.3.3.1 Building Scientific and Technical Coordination Mechanism

The CWRC, together with the water resources departments of the provinces, autonomous regions and municipalities where the key reservoirs are located and reservoir management units, formulates the joint operation scheme of the key reservoirs in the Yangtze River Basin based on the operation demands, scientific ground and stakeholder consultations, then submits it to the Ministry of Water Resources for approval. According to the approved scheme, the above-mentioned water resources departments and reservoir management units work out the annual operation plans of their jurisdictional reservoirs, reporting or submitting them to the CWRC for approval or just notification. Referring to the approved annual operation plans and with the aid of hydrological forecast and monitoring, the CWRC issues dispatching commands after consulting on rolling basis with the water resources departments and reservoir management units, and the reservoir management units implement the commands accordingly.

10.3.3.2 Setting up Rational Interests Balancing Mechanism

The joint operation of key reservoirs may entail internal readjustment of the economic benefits, social benefits and eco-environmental benefits of the reservoir itself as well as the redistribution of interests among the reservoirs. Therefore, it is necessary to develop corresponding compensation mechanism to balance the interests between the State and corporation, and among different corporations. It is also expected to set up joint operation management fund to compensate the reservoirs that suffer economic loss, subsidize operation scheme formulation, finance the development of the comprehensive management information system and supervisory management.

10.3.3.3 Establish Information Sharing Mechanism

The joint operation of the group of key reservoirs requires comprehensive and in-time monitoring data and the sharing of them. It supposes to improve the comprehensive data collection networks on the basis of the existing Yangtze River Basin hydrological regime monitoring station network and such network of the key reservoirs, and adds in necessary hydrological, water quality and hydro-ecological monitoring stations and integrates the monitoring stations run by relevant departments, local governments and the reservoir management units. It is necessary to integrate the information sourced from various departments, local governments and reservoir management units through an information sharing platform, so as to realize sharing of the monitoring, dispatching and operational information.

10.3.3.4 Establishing Broad-Based Public Participation Mechanism

The joint operation of key reservoirs needs the broad-based public participation. It supposes to make open to the public and stakeholders, as they have the due rights to know, the information concerning reservoir operations. The related working rules should be established to allow the public to participate in the whole process of reservoir operation. Moreover, supervision mechanisms shall be effecuated to empower the public to supervise if their feedback opinions are properly considered and implemented through establishing such supervision mechanism.

10.3.4 Strengthening Technical Support

The management of joint operation of the key reservoirs in the Yangtze River Basin, having widespread influences and involving complicated technologies, requires strong technical supports from hydrological monitoring and forecasting, scheme and scenario plan making, as well as enhanced hydro informatics.

10.3.4.1 Hydrological Monitoring and Forecasting

The comprehensive monitoring system of the basin should be improved constantly. The information of the basin-wide water works can be shared through building up an information platform that allows information collection and storage, information service and supporting applications. Efforts should be made to probe deep into the methodologies on rainfall forecasts in medium and short term, including synoptic forecast, numerical forecast, remote sensing forecast and comprehensive forecast. Enhance the studies on the methodologies of ensemble forecast and probability-based forecast to explore the possibility of quantifying the risk and probability of the real-time forecast. It is necessary to strengthen studies on the model and methodology on water quantity forecast in medium and long term. The accuracy of hydrological forecast can be improved and the lead time can be extended through the above measures.

10.3.4.2 Scheme and Scenario Plan

The objectives in integrated operation of the group of key reservoirs include flood control, water supply, water quality, hydro-ecology, hydro-power generation and navigation. Based on detailed study on the operation mode for a particular individual objective of each of these reservoirs, the multi-objective integrated regulation and management of each of the reservoirs can be worked out through integration and balance of all the objectives. Models on multi-objective operation of group reservoirs should be set up and continuously improved by doing simulations and modifying performances under different hydro-meteorological scenarios. Furthermore, to improve the joint operation scheme, focus should also be put on determining, for each of the reservoirs, the flood control limit water level and flood control operation mode in wet season, required water release and reservoir impounding operation mode during fill-up period, operation of active storage and required water release to downstream in water supply period, limits on the outflow and draw-down operation prior to flood season.

10.3.4.3 Informatization Capability

Given the river basin management mandates and the demands for a comprehensive system to integrate operations of all the water works, the existing infrastructure of hydro informatics in the Yangtze River Basin shall further incorporate all the operational systems related to reservoir regulation currently in service, and emphasize the application in water sector of advanced IT technologies such as big data, cloud computing, internet, 3S and VR, etc. By means of add-in and upgrade, it supposes to build up a highly integrated basin-level water works joint operation system featured by multi-sources of data, big data platform, and uniform technical architecture. Such system, empowered with strong capability in information sharing, integration of applications, and operational synergy, will play a critical role in decision support with respect to warning, forecast and dispatching, for better public service and river basin management.

10.3.5 Enhancing Supervision and Management

Enforcing the joint operation and management of the key reservoirs in the Yangtze River Basin require powerful and effective supervision. Clear legal definitions on the competence, responsibility and accountability for both supervisors and the operators involved in the process, over which an effective supervision and management mechanism can grow matured progressively and serve as safeguards to ensure the effective implementation of such joint operations. Meanwhile, the water sector regulatory decrees and directives need to be enacted and enforced by the Ministry of Water Resources, such as the decree on the supervision and management of reservoirs on flood control limit water level, and the directive on performing hydraulic works operations.

10.3.6 Strengthening Researches

10.3.6.1 Joint Operation of Water Works for Flood Control

Studies should be strengthened on the storage volume compensation between different reservoirs in joint operation for flood control. The current joint operation for flood control with the Three Gorges Reservoir as the core should keep optimizing to enhance flood control capability. It is also urgent to establish the modelling for dynamically simulating the operating water levels of the group of reservoirs, for the purpose of more efficiently utilizing floodwater as resources. In order to improve the overall basin-level flood control capacity, researches should also focus on developing the integrated flood control operation mode for all the structural components to work together. For instance, the operation mode taking into consideration of the designated detention basins and those previously encroached floodplains and polders which already have been returned to flood-ways or to flood detention basins, and the Three Gorges Reservoir, in addition to integration of the large-size drainage pumping station on the middle-lower reaches. It is also vital to research in more depth on the scheme and scenario plan on flood control on the main tributaries, such as the Yalongjiang River, Minjiang River and Jialingjia River. Hence, such scheme and scenario plan could provide better guidance for real time flood regulation.

10.3.6.2 Integrated Regulation on Water Quantity Allocation

An integrated water resources allocation and regulation system needs to be developed in the Basin in order to integrate water resources allocations along the river, and intra-basin and inter-basin water transfers. Operations of water diversion and transfer projects should be subjected to a unified management system, and the water withdrawals and water uses of the South-to-North Water Transfers should be put under strict supervision. There is a need to speed up the paces of establishing rules and mechanisms for the rational allocation of water resources through trans-basin water diversion and transfer, the safeguards of eco-environment water use, and for the ecological compensations. It is designed to secure basin level water supply through highly efficient integrated water resources allocation and regulation, as such, water can be supplemented where and when necessary to relieve the droughts. The balance between water export and water import areas, and between resources-environment carrying capacity and socio-economic development has to be properly managed. The operation plan of water works in the middle and lower Basin for water supply in extremely dry years is needed.

10.3.6.3 Eco-friendly Operation of Water Works

The demands for eco-friendly operations of the upper Yangtze River cascade reservoirs need to be identified through in-depth investigations into the intrinsic interaction between the ecosystem's demands and the operations of group reservoirs. To prevent and control algal blooms in the reservoirs through joint operation of the group reservoirs is a subject of research study, in which identifying a selective set of parameters and their quantification are a focus, and the research output shall guide how to operate the reservoirs to curb algal blooms in the reservoirs and tributaries in different seasons and at different water levels. Researches are also needed on how to manoeuvre the upper Basin reservoirs to help the important aquatic species to survive through the critical stage of their life cycle, particularly the researches shall ascertain the tempo-spatial distribution of the Chinese Sturgeons and the four Yangtze River carps, their natural breeding situation, their spawning ground conditions including hydrographic and water temperature requirements, as well as their requirements for flow regime and water temperature in their critical developmental process. Experience and findings gained from previous experimental eco-friendly operations shall be documented and reviewed to optimize the ecological operation of the reservoirs.

10.3.6.4 Multi-objective Joint Operation

There is a need of in-depth researches on theories of decision making under risk and risk assessment in joint operation of the multi-objective reservoirs, with systematic indicators of risk assessment to be established. Post-evaluation to identify the effect of the reservoir operations on ecosystem and the compensation mechanism of interest losses resulting from the risk-based operation need to be carried out. To conduct studies on modelling the coordinated operation to meet multiple objectives of the reservoirs will lead to the scheme of joint operation of the multi-objective reservoirs in the Yangtze River Basin. For emergency response to major accidents threatening water security, the techniques of emergency operation of the group reservoirs in the upper Yangtze River Basin need to be developed, in which particular importance is to define clearly the emergency operation's principles, focal points, procedures and synergic mechanism.

Chapter 11
Water Resources Protection

11.1 Development Process and Achievements

In 1972, the United Nations Conference on the Human Environment was held in Stockholm, Sweden. Since then, China's top leadership has raised the awareness that China also face serious environmental problems like many other countries in the world.

Against this background, China's first National Conference on Environmental Protection was held in Beijing in August 1973. An official document was drawn up, "Several regulations on environmental protection and improvement (for trial implementation)", which clearly required the establishment of environmental protection and management organizations at basin level for all major rivers and lakes in China.

In January 1976, under the approval of the State Council Environmental Protection Leading Group and the Ministry of Water Resources and Electric Power, the Yangtze River Water Resources Protection Bureau was founded, which later named as the Yangtze River Basin Water Resources Protection Bureau (hereinafter referred to as "YBWPB"). It was an epoch-making event, as the YBWPB was China's first institutional establishment for basin water resources protection, and its founding marked the official starting point for the protection of water resources in the Yangtze River Basin.

11.1.1 Process

There are three development stages regarding the water resources protection for the Yangtze River: start-up stage (prior to 1990), fledging stage (1990–2000), and flouring stage (since 2000).

© Changjiang Press (Wuhan) Co., Ltd. 2022

Changjiang Water Resources Commission of Ministry of Water Resources of the People's Republic of China, *Report on Yangtze River Rehabilitation and Protection 2019*, https://doi.org/10.1007/978-981-16-4927-1_11

11.1.1.1 Start-up Stage (Prior to 1990)

At this stage, there were several key tasks of water resources protection in the Yangtze River: mapping pollutant discharge conditions of the Yangtze River system, formulating the Yangtze River water resources protection planning, establishing rules and regulations for water resources protection, building a water quality monitoring network, monitoring water quality and pollutant source, and carrying out research on environmental capacity of water bodies and laws of dilution and natural purification of pollutants.

(A) **Organizing the Formulation of Water Resources Protection Planning for Main Stream of Yangtze River**

In the mid-1980s, the Ministry of Water Resources and Electric Power, together with the Ministry of Urban–Rural Construction and Environmental Protection, organized the formulation of the water resources protection planning for seven major basins in China for the first time. One of them was Report on Water Resources Protection Planning of Mainstream Yangtze, completed in 1987. It was the first relatively comprehensive planning of its kind for the Yangtze River Basin, which served as an important foundation for relative work.

(B) **Initiating Monitoring in the Yangtze River Basin**

Among the first in China, the water quality monitoring network of the Yangtze River system was established, and the Suggestions on Water Quality Monitoring Network and Monitoring Planning of the Yangtze River System was issued. In 1979, the Planning of Water Quality Monitoring Network in the Yangtze River Basin was completed, which provided regulations for monitoring network set-up principle, network planning and monitoring technology, and data processing. Also based on the requirements of this Planning, water quality monitoring was carried out for a variety of water bodies including rivers, lakes and reservoirs, and a brief understanding was built on surface water quality condition in China at the time.

(C) **Establishing Management Regulation System**

In 1978, the Yangtze River Water Resources Protection Bureau, in cooperation with relevant experts, carried out extensive investigations and drafted the Regulations on Water Resources Protection in the Yangtze River, which was reported to the Ministry of Water Resources and Electric Power and the State Council Environmental Protection Leading Group. In 1984, the Ministry and the Leading Group co-held the Work Conference on Yangtze River Water Resources Protection, and approved two documents, Some Provisions on Yangtze River Water Resources Protection, and Regulations on Water Quality Monitoring Network in the Mainstream of the Yangtze River. These two documents were jointly issued by the Ministry of Water Resources and Electric Power and the National Environmental Protection Bureau for implementation among all river basin provinces, autonomous regions and municipalities.

(D) **Carrying out Scientific Research**

In 1978, the report on Status of water resources pollution in the Yangtze River was completed. This report proposed some key concepts for the first time in China, including ratio of sewage discharge to river runoff, near-shore pollution zone, and pollution load. It also proposed countermeasures for water resources protection in the Yangtze River. During this stage, a research institute and a monitoring center for water resources protection to provide technical support to relative work have been established in the country for the first time. Meanwhile, the first development outline for water resources protection research has been formulated, in which four key research topics were listed, including water quality monitoring system in the Yangtze River, pollutant migration and transformation rules and water environmental capacity in the Yangtze River, comprehensive prevention and control of pollution in the Yangtze River, and the environmental impact of large scale water conservancy projects.

11.1.1.2 Fledging Stage (1990–2000)

From late-1980s to 1990s, with the promulgation and implementation of China's Environmental Protection Law and Water Law, we took a further step in clarifying responsibility, defining division of work and strengthening management for water resources protection in the basin. The most distinctive progress made during this stage included the implementation of water functional zoning division, and the re-formulation of water resources protection planning for the Yangtze River Basin.

(A) **Continuing Advancing Top-Level Design for Planning**

Staring from the year 2000, the Ministry of Water Resources carried out nation-wide water resources protection planning work and completed the Report on Water Resources Protection Planning in the Yangtze River by the end of 2001. For the first time, water functional zoning was used as the basic unit of the planning. Based on the calculation of pollutant carrying capacity of water functional zoning, the Report proposed feasible countermeasures for controlling total pollutant discharge and pollution reduction volume. This planning was regarded as the first whole basin water resources protection planning, comparing with the planning for mainstream Yangtze in the 1980s, and great progress was made in planning guidelines and methodology.

(B) **Consolidating Water Resources Protection Management**

In 1994, the Report on Investigation and Evaluation of Discharge Outlets in the Mainstream Yangtze was completed, which provided an overview of the condition of discharge outlets in the mainstream of the Yangtze River. In January 1998, YBWPB hosted the Yangtze River Water Resources Protection Conference in Huangshi,

Hubei, and the initiative of "Promoting sustainable utilization of water resources for the protection of the Yangtze River" was released, suggesting that all enterprises in the basin work together to preserve a clean Yangtze River for the new century to come.

A feasibility study was carried out in 1997, regarding legislation on water resources protection in the Yangtze River. As a result, a general report and four special reports on the feasibility of legislation on the Measures for Water Resources Protection in the Yangtze River came out, which provided theoretical support for the legislative work later on.

From January 1999, investigation and research was carried out jointly by YBWPB and Water Environment Evaluation Center of MWR, on drinking water sources monitoring technology and the compilation of a ten-day report.

Since 2000, all provinces (cities and districts) in the Yangtze River Basin have been mobilized into the tasks of defining water functional zoning, and the Report on Water Functional Zoning of the Yangtze River was completed the following year.

(C) Enhancing Monitoring Network and Capacity Building

In 1994, water environment monitoring network in the Yangtze River Basin has been adjusted based on years of experience. In May of the same year, the Work Conference for Water Environment Monitoring in the Basin was held in Shanghai, and the document Measures for Water Environment Monitoring Network Management in the Yangtze River Basin was deliberated and passed. During the period, YBWPB completed several important reports, including, Report on Adjustment of Water Environment Monitoring Network in the Yangtze River Basin, Suggestions on Adjusting Water Environment Monitoring Stations and Monitoring Cross-sections in the Mainstream of the Yangtze River, Report on the Ninth Five-year Planning of Water Environment Monitoring Network in the Yangtze River Basin, and Report on the Construction Planning of Water Environment Monitoring Stations at Provincial Boundaries in the Yangtze River Basin. These reports positively supported the upgrading of water resources protection monitoring network and the improvement of monitoring capacity.

(D) Carrying out Scientific Research in Key Areas and Major Regions

In 1994, the report of Study of Water Environment Condition of Near-Shore Areas in Major Cities of Mainstream Yangtze was completed, which evaluated the water body condition of near-shore areas for the first time in a comprehensive and objective manner. Since 1996, YBWPB has undertaken the special subject on Countermeasures Research for Water Pollutant Control of the Three Gorges Reservoir. Based on investigation and analysis of pollutant sources and water quality of the Three Gorges Reservoir, the research determined the water functional zoning of the reservoir and water quality protection goals, drafted the calculation procedure of water environment capacity, and finally put forward countermeasures including total discharge limit, load distribution, pollution controlling standards, and water quality management measures.

11.1.1.3 Flouring Stage (Since 2000)

Entering into the twenty-first century, the negative ecological and environmental impacts of unreasonable use of water resources are gradually appearing, raising growing attention to the Chinese government and the public. As a result, water resources protection in the basin is being further strengthened, with two major tasks, enhancing basin water resource protection planning based on water functional zoning, and improving supervision and management of water functional zoning and sewage outlets.

(A) **Formulating a Complete Planning System**

In 2005, CWRC launched a new round of work for the master plan in the Yangtze River Basin, in which CWRC emphasized the improvement of four systems, namely, flood prevention and disaster mitigation system, water resources comprehensive development system, water resources and ecological environment protection system, and integrated river basin management system. It was the first time that the protection of water resources and ecological environment was highlighted in the master plan of the basin and stressed a shift from ex-post remedy to prevention.

This period also witnessed more emphasis from the central leadership. In 2011, clear requirements for enhancing water resources protection were put forward in the central government's First Document of the Year, as well as the Central Work Conference on Water Resources. In the beginning of 2012, the State Council of China issued the Opinions on the Implementation of the Strictest Possible Systems for Water Resource Management by the State Council. And in 2014, the General Secretary XI Jinping put forward the guidance for water resources governance as "prioritizing water conservation, balancing development with spatial distribution of water resources, practicing systematic governance and achieving government-market synergy".

To strengthen the protection of water resources, a new round of planning was started in 2012, under the arrangement of the MWR and the CWRC. The newest Water Resources Protection Planning in the Yangtze River Basin served as the top-level design in the new era, and its content was greatly expanded, from simply focusing on water quality, to an integration of water quality, quantity and ecology protection. At this stage, the water resources protection planning system has been gradually improved, which provided an important basis for the following period.

(B) **Enhancing Supervision and Management**

In 2002, the Opinions on Improving Water Resources Protection of the Yangtze River in the Near Future was completed, which provided guidance to relative work and proposed key tasks for the period. In the same year, the new Water Law was introduced. Against this background, CWRC actively participated in the formulation and revision of regulations, including the Administrative Measures for Water Functional Zoning, and the Administrative Measures for Supervision and Management of Sewage Outlets. For the better implementation of these documents, CWRC drafted and issued the Implementation Detailed Rules for Supervision and Management of Sewage Outlets by the Changjiang Water Resources Commission, and the Measures for the Establishment and Acceptance of Sewage Outlets by the Changjiang Water Resources Commission.

Since 2000, water functional zoning supervision and management system has been improved constantly, with the introduction of a sound monitoring and evaluation system. Monitoring, supervision and inspection of buffer zoning at provincial boundaries beyond standards have been strengthened, and supervision management on trans-provincial rivers and lakes has been further enhanced, through the construction of buffer zoning signs at provincial boundaries. Based on improved supervision and management on major water functional zoning, the integration of regional and basin management has been facilitated. Annual inspection and evaluation have been organized on major drinking water source areas, and positive results have been achieved for the goals of "ensuring water quantity and quality and improving monitoring and a sound management system."

(C) Establishing and Strengthening Monitoring System

In 2000, the Yangtze River Basin Water Environment Monitoring Center, under the YBWPB, organized provinces within the basin to jointly complete the Planning of Yangtze River Water Quality Monitoring. This Planning combined the monitoring network planning at both basin and local level and drew up the overall planning for monitoring stations including surface water, ground water, and atmospheric precipitation, which was the first comprehensive planning for water quality monitoring in the Yangtze River Basin. For the first time, capacity building, technical training and information system construction were included in the Planning and were considered in a holistic manner together with station planning, which laid a solid foundation for improving monitoring capacity, and accelerating the development of water quality monitoring in a standardized, scientific, and modernized direction. The monitoring system for water resources protection in the basin has been increasingly improved after hard work of nearly two decades, which has become an essential part, like "eyes and ears," for the rehabilitation and protection of the Yangtze River. The layout of stations and monitoring capacity has remained at the forefront of the industry.

(D) Carrying out Extensive Scientific Research

After 2000, scientific research in the field of water resources protection has gradually expanded to include comprehensive research on pollution carrying capacity and total discharge limit, health assessment of rivers and lakes, monitoring technology, early warning and forecasting, ecological compensation, legislation, and relations between water resources development and protection. In 2004, the report of Opinions on Carrying Capacity and Total Discharge Limit of the Three Gorges Reservoir was completed, which laid a technical foundation for drafting the code of calculating pollutant carrying capacity of water bodies later. From 2004 to 2006, an innovation project of Research on the Relations between Water Resources Development and Ecological Environment Protection in the Yangtze River Basin was led by YBWPB, which later won the First Prize of Dayu Water Conservancy Science and Technology Prize in 2009, one of the top honors in the field in China. In 2010, the program for health assessment of major rivers and lakes in China was launched by the MWR. According to the requirement of the MWR, CWRC established the leading group and technical group for rivers and lakes health assessment in the Yangtze River Basin, and actively facilitated related provinces and regions within the river basin to accomplish two phases of pilot work. In 2015, CWRC organized researches and demonstration studies on legislation of water resources protection regulations in Guizhou Province, which later served as a key basis for the promulgation of Water Resources Protection Regulation in Guizhou Province in 2017.

11.1.2 Achievements

The past four decades have witnessed the development course of water resources protection of the Yangtze River. Through practical work, we've made extensive exploration and accumulated experience, based on which water resources protection theory has been constantly enriched and improved. As a result, significant achievements have been made in the building of four systems, namely, water resources protection planning system, supervision and management system, monitoring system, and scientific research supporting system.

11.1.2.1 Water Resources Protection Planning System

A planning system of "paying equal attention to water quality, quantity and ecology, making overall coordination while highlighting key points, strengthening supervision and management, and employing both engineering and non-engineering measures", has been gradually built from scratch, greatly supporting the development of the water resources protection system in the basin. The breakthrough made in the engineering measures system provided an especially effective measure for practical work. After years of exploration and efforts, a series of planning code have been established, including Code for Calculation of Pollutant Carrying Capacity in Water Bodies, Water Functional Zoning Classification Standards, and Code for Water Resources Protection Planning, which have made remarkable contribution to the standardization of water resources protection planning.

11.1.2.2 Water Resources Protection Supervision and Management System

Through long-term exploration and practice, the water functional zoning management system in the Yangtze River Basin has been gradually improved. Several key tasks have been accomplished. First, the classification of water functional zoning of key sections within the basin has been completed. Second, the pollutant carrying capacity of key water functional zoning within the basin has been verified, a phased plan for total discharge control has been established, and the annual target rates for water functional zoning have been set. Third, the provincial buffer zones management has been strengthened, through the reviewing of provincial (buffer zone) water quality monitoring cross-sections in the Yangtze River Basin and the southwestern rivers. Fourth, water functional zoning monitoring has been organized, and support has been offered to the assessment of water functional zoning regarding the pollutant control red-line policy. The above works played a key role in enhancing supervision and management of water functional zoning and implementing the strictest possible systems for water resource management mechanism.

11.1.2.3 Water Resources Protection Monitoring System

Two comprehensive systems have been built for enhancing monitoring during the past years. The first is a water environmental monitoring network, with a basin-level monitoring center as headquarter, the basin sub-centers and provincial-level centers as backbone, and the city-level centers as basis. The second is a water resources protection comprehensive monitoring system, combining regular manual and automatic monitoring, fixed point and mobile patrol monitoring, and fixed time and real-time monitoring.

11.1.2.4 Water Resources Protection Scientific Research Supporting System

Over the past decades, pioneering results have been achieved in scientific research on the protection of water resources in the Yangtze River, which won several national and provincial awards. Studies have been carried out on the basic theory, comprehensive management, monitoring technology and applied technology of water resources protection, such as water chemical characteristics, pollutant migration and transformation rules, water quality monitoring methods, water quality mathematical model, pollutant carrying capacity and discharge limit control technology of water bodies, ecological compensation in regional areas, river health management, and etc. The researches mentioned above provided solid support for better protecting water resources and realizing management functions.

Summary

With all these efforts, the water environmental condition of major rivers in the basin has been improved in a comprehensive way. The water quality standard-reaching-rate of water functional zoning increased to 81.3%. Drinking water security in cities has improved significantly, as the security guarantee rate for city centralized drinking water sources reached almost 100%. The water resources protection monitoring capability has been comprehensively improved in the following areas: the full coverage monitoring of provincial boundary sections of the Yangtze River Economic Belt has been realized; the monitoring coverage of water function areas in important rivers and lakes has reached over 97%; and more than 90% of 215 major drinking water sources in China have been monitored. A central laboratory under the basin monitoring center, which is the largest in the region, has been built, with complete and state-of-the-art hydrological and water quality monitoring device, mobile monitoring labs, and automatic water quality monitoring stations. Generally speaking, the basic water resources protection system and mechanism, combining management at both basin and regional level, have been formed.

11.2 Status and Problems

11.2.1 Water Environment Status of Rivers/Lakes

11.2.1.1 Pollution Sources

According to the 2017 Public reports on water resources of the Yangtze River Basin and southwest rivers of China, the total annual discharge of waste water in the Yangtze River Basin (excluding the Taihu Lake system) was 28.89 billion tons (not including the 20.98 billion tons of once-through cooling water of thermal power plants and pit drainage), of which the domestic waste water was 13.06 billion tons (including 4.97 billion tons from the tertiary industry and construction industry), accounting for 45.2%; the industrial waste water was 15.83 billion tons, accounting for 54.8%. Regarding the spatial distribution, discharge was mainly concentrated in the following areas classified as Secondary water resources zoning: the Dongting Lake and its tributaries, mainstream of the Yangtze River downstream of Hukou, and Poyang Lake system, which were in the administrative divisions of Hubei Province, Hunan Province, Jiangxi Province and Sichuan Province. As Fig. 11.1 demonstrates, the annual total discharge of waste water and annual domestic waste water discharge in the Yangtze River Basin increased from 2014 to 2017, while the industrial discharge showed a slightly decreasing tendency.

11.2.1.2 Water Quality

According to the 2017 Public reports on water resources of the Yangtze River Basin and southwest rivers of China, the water quality of the Yangtze River was in relatively good condition, with 83.9% of assessed river length reaching Grade I, II, or III conditions, and the rest 16.1% worse than Grade III. Among the 164 provincial cross-sections, 89.6% reached water quality of Grade I, II and III all year round. About 14.8% of 61 lakes and 81.8% of 362 reservoirs evaluated were classified as category I, II and III, while 85.2% of these lakes and 33.4% of these reservoirs

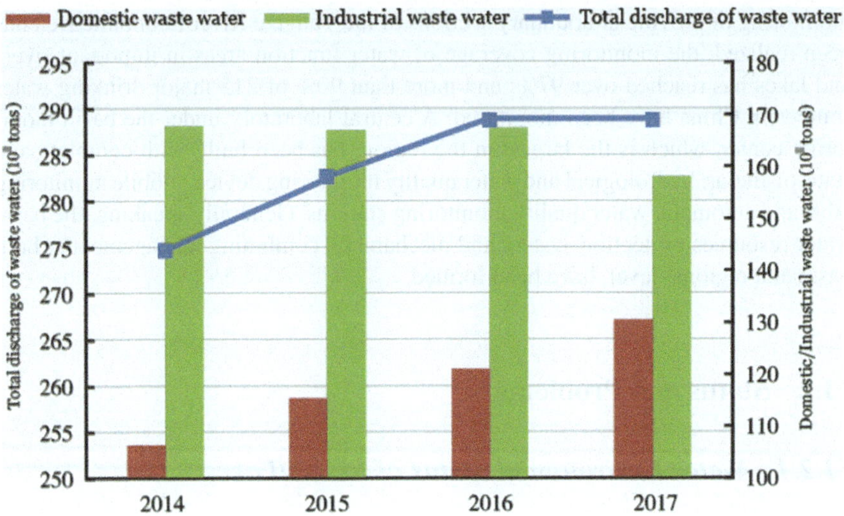

Fig. 11.1 The change of waste water discharge in the Yangtze River Basin

were in mild or minor conditions of eutrophication. Among the 1261 major water functional zones listed in the Water Functional Zoning of Major Rivers and Lakes in China (2011–2030) approved by the State Council, the standard compliant rate for full-index and two-index (CODMn, NH_3–N) standard were 78.0% and 93.1%, respectively. Among the 515 evaluated water sources, 73.2% were up to the standard for the whole year, and 90.3% reached a standard-compliant rate of 80% or above.

11.2.1.3 Drinking Water Source Areas

According to the results of the assessment of the safety standards of important national drinking water sources in the Yangtze River Basin in 2018, the overall condition in 2017 was classified as Excellent, with 127 source areas (58%) rated as Excellent, 66 (30.4%) rated as Good, and 21 (9.7%) rated as Satisfactory.

11.2.2 Problems

Due to the unremitting efforts in recent years, water resources protection in the Yangtze River Basin has been strengthened. The trend of increasing water pollution has been effectively controlled, water quality of mainstream Yangtze has been kept in a good condition, and water safety management has been improved. Meanwhile, with the rapid socioeconomic development in the basin, the waste water discharge volume is increasing year by year, though the growth rate has been slowed down.

Water pollutant remains a serious problem for certain sections, tributaries, lakes and reservoirs. The situation of water resources protection in the Yangtze River Basin still remains serious.

11.2.2.1 Pollution Remains Serious in Near-Shore Waters and Some Tributaries, and Eutrophication Is Still Extending

Generally speaking, water quality in the Yangtze River Basin is in a good condition at present, but certain sections of cities in the mainstream and certain part of tributary water bodies are still facing pollution to varying degrees. With the acceleration of urbanization and a growing demand and supply of water resources, pollution will be even severe if not taking effective measures. Some tributaries of the Yangtze River have been polluted in different degrees, for tributaries such as Minjiang River and Tuojiang River, pollution is serious in certain sections; and for other rivers such as Xiangjiang River, measures for tackling heavy metal pollution are not yet in place.

Another problem is eutrophication. In recent years, due to the human activities for living and production, the ecological environment of lakes and wetlands has been affected. Because of the excessive use of chemical fertilizers and pesticides, non-point source pollution has become increasingly prominent. And as the pollution of nitrogen and phosphorus is still increasing, eutrophication remains a serious problem. For lakes listed by the government as special focus for eutrophication control, such as Chaohu Lake and Dianchi Lake, the situation has not been effectively controlled; water quality of the Dongting Lake is deteriorating and eutrophication is expected to be observed; the Poyang Lake is in the mesotrophic status; and some tributaries in the Three Gorges Reservoir are in a state of mild eutrophication.

11.2.2.2 Ensuring Drinking Water Safety in Urban and Rural Areas in River Basin Remains Difficult Task

In the past few years, the central and local governments have stepped up their efforts to ensure drinking water safety in both urban and rural areas by adopting a series of engineering and management measures and have addressed some related problems. However, much work still needs to be done in order to meet the requirements for improving water safety. Meanwhile, rapid socioeconomic development, coupled with an increasing population and acceleration of urbanization, has generated more threats for water safety management.

First, urban water supply shortage is increasing. Due to the growing urbanization rate and water consumption, supply becomes insufficient for lack of, either enough water resources, or clean and potable water, or water supply infrastructure, causing difficulties for people's life and harming the sustainable development of the cities. Second, water sources are being polluted in some cities and towns in the basin, especially with increasing organic pollution, which poses threats to people's life and health. Third, plants and factories with high pollution risks are densely constructed along the river, intensifying the pressure to prevent major water pollution incidents.

11.2.2.3 Ecological Security Faces Greater Threat with Damaged Water Ecosystem

In recent years, achievements have been made in the protection of aquatic habitats and biodiversity. But there is still a long way to go, due to the negative impacts of unreasonable development and utilization of the ecological environment for a long time.

First, the negative impacts of river basin development on ecological environment have gradually revealed, leading to a reduced natural connectivity of rivers and lakes, and serious threat to biodiversity and integrity caused by the changes of habitat. Second, the construction of diversion-type hydropower stations and inter-basin water diversion projects has caused a sharp reduce of flow, or even seasonal dry up in the river, resulting in insufficient ecological water use, threatening river health. Third, the task of preventing and controlling soil erosion remains arduous, as damages still exist.

11.2.2.4 Monitoring Capacity for River Basin Water Resources Protection Requires Further Improvement

Monitoring capacity of water functional zoning needs to be further enhanced. The full coverage of monitoring is not yet achieved in the basin, and we are facing problems such as insufficient monitoring frequency and incomplete monitoring indicators, which make it hard to meet the requirements for the most stringent water resources management.

Monitoring of sewage outlets is still lagging behind. Comprehensive and systematic monitoring has not been carried out yet. It is still difficult to accurately calculate the total discharge volume of pollutants into rivers (lakes), because of insufficient monitoring quantity, as well as few parameters and low frequency.

Water ecological monitoring is developing rapidly, but still in the elementary stage. Aquatic species monitoring has been conducted with a pilot monitoring on algae, while monitoring on other aquatic species and habitat still lagging behind.

All-indexes-monitoring on drinking water source areas has been carried out, but not in a comprehensive and systematical way. Currently, there are 80 special parameters listed for centralized drinking water source areas monitoring, but regular monitoring has not been implemented due to insufficient monitoring capacity and funds.

Meanwhile, there are still gaps between the requirements of water resources protection, and status quo for monitoring of sewage outlets, water ecological monitoring, emergency monitoring, and informatization level. More efforts are required urgently in enhancing monitoring, improving capacity of monitoring technical personnel, and applying new technology and equipment.

11.2.2.5 Water Resources Protection Management Needs to Be Further Enhanced

There is still room for improvement when it comes to management mechanism and system, and supervision and monitoring capacity for water resources protection, to properly support socioeconomic development in the basin.

First, the legal framework for water resources protection needs to be enhanced. There is an absence of comprehensive protection regulation at basin level, while the cohesion and coordination between existing laws and regulations are weak. Most stipulations for water resources protection are dispersedly embodied in the Water Law and Water Pollution Prevention Law. As a result, overlapping and unclear division of responsibilities can be found among sectors and industries, and between basins and regions.

Second, there is a great urgency in establishing a cross-sectoral and trans-regional coordination mechanism. The existing water resources protection coordination mechanism, which aims to integrate basin and regional management, is yet to be upgraded, and the joint coordination among sectors or regions is still inefficient. Problems such as dysfunctional consultation and dispute settlement mechanism, poor access to information, and insufficient information sharing, still exist.

Third, the long-term supervision and management mechanism for water functional zoning and sewage outlet needs to be enhanced. Some difficulties still exist in the implementation of pollutant carrying capacity and total discharge limit of water functional zoning in the basin, which requires refining the existing regulations and establishing the adjustment management mechanism for discharge limit. The classified and tiered management of sewage outlets needs to be further improved at both basin and regional levels.

Fourth, the emergency management of pollution incidents needs to be strengthened. Currently, industrial parks are densely distributed along the river, with a large handling capacity of toxic and hazardous chemicals. Moreover, it has become a accelerated tendency for the heavy chemical works to move to the middle and upper reaches. All these increase the risk of pollution incidents, but the coping capacity for emergent water pollution incidents remains weak.

11.3 Prospects on Development

As we look into the future, unprecedented opportunities is presented in water resources protection in the Yangtze River Basin, with the implementation of the strictest possible systems for water resource management, the river and lake chief system, the Yangtze River Economic Belt development strategy, and the idea put forward by Xi Jinping of "promoting well-coordinated environmental conservation and avoiding excessive development in the Yangtze River Basin". China has sped up the institutional reform and has established the basic framework for developing an ecological civilization, providing a positive impetus and momentum to tackle obstacles in water ecological environment protection and management. Moreover, the implementation of river and lake chief system provides effective measures to solve water environmental problems, improve water ecological environment, and preserve healthy rivers and lakes.

Against this background, we should seize the four major strategic opportunities in advancing water resources protection in the basin. Regarding pollution control, based on water functional zoning management, we shall implement strict control on total pollutant discharge into the rivers, further promote comprehensive regulation for main river reaches and key tributaries, and enhance eutrophication control of lakes and reservoirs, so as to bring discharge volume under control step by step. Regarding drinking water safety, we shall carry out comprehensive pollution control program for water source areas, to ensure a benign ecosystem of water source areas and better water quality. Regarding water ecosystem protection, guided by the goal of ensuring a healthy lake–river system, we shall promote the sustainable development of water resources, and shall ensure the ecological flow of key sections of mainstream and tributaries, by improving ecological scheduling of water projects. Through coordination of concerted efforts, enhanced fundamental works and capacity, we shall promote water resources protection and management to a higher level in a comprehensive manner and improve the five key systems, namely, water resources protection regulation and mechanism system, planning system, management system, supervision and monitoring system, and scientific research supporting system.

There are several key points of water resources protection in the Yangtze River Basin at present and in the near future.

11.3.1 Advancing Institutional Construction and Legislative Process for Water Resources Protection

With the implementation of river chief system, efforts will be made in establishing mechanisms, including, the coordination mechanism for planning of water resources protection and water pollution control in key areas such as the upper Yangtze River hydropower development area, the Three Gorges Reservoir area, the Danjiangkou Reservoir area, and the middle and lower reaches; the trans-regional and cross-sectoral mechanism for water resources management and protection; and a clear responsibility for ensuring ecological flow of important control sections in the basin. We shall speed up the construction of supervision and management mechanism for hydropower projects in key areas and key projects in the basin, joint operation of inter-basin water diversion projects, and ecological flow maintenance. Ecological compensation for areas including the Three Gorges Reservoir, the Han River basin, the Chishui River basin, and major drinking water source areas, will be further emphasized as a priority. Information disclosure and public participation will be further enhanced.

We will push forward the legislation and implementation of the Yangtze River Protection Law, and enhance accordingly the supporting regulations at local level. Step by step, we will establish a relatively complete river basin water resources protection regulation system, with laws such as the Water Law and Water Pollution Control Law as the core, and administrative regulations, departmental rules and water-related laws at local level as support. To this end, we will speed up the formulation of local water resources protection regulations, as well as management regulations on water functional zoning, sewage outlets, and drinking water source area protection, which will contribute to the enhancement of water resources protection regulations system in the basin.

11.3.2 Improving Top-Level Design and Pushing the Follow-up Implementation

Efforts will be made to improve the planning system. We will accelerate the formulation of special and subject plans for key branches such as Jialingjiang River, Min River, Han River and Xiang River, key lakes such as Poyang Lake and Dongting Lake, and key areas such as river source areas and estuary areas of the Yangtze River. Our goal is to build an improved water resources protection planning system, which integrates basin and regional management, and combines master and special plans. At the same time, we will facilitate the implementation of the plans in an orderly manner and carry out evaluation in due time. We will promote the water resources protection pilot projects in key areas, and gradually establish and improve the system of water resources protection projects in the basin.

11.3.3 Enforcing Appraisal Against the Red Line and Supervision

We will strictly review and ratify the limit red lines of pollutant assimilation in key water function areas and improve the system for evaluating and assessing water quality in these areas. Supervision on sewage outlet management will be further enhanced. Regarding important drinking water sources in the basin, we will strengthen efforts to ensure the safety and standard reaching rate, and to promote the management of ecological flows in important sections, so as to improve the water resources protection and management system. We will push ahead the establishment of a routine linkage mechanism for coordination in water administrative law enforcement at basin and regional levels, and a trans-regional joint law enforcement mechanism combining basin and regional level enforcement, of a clear division of responsibility and well-coordinated operation. We will build cross-sectoral joint law enforcement cooperation mechanism and will actively explore joint actions in law enforcement between water administrative sectors and environmental protection sectors by making unified plans and taking joint actions, to intensify inspections and form a closely-coordinated cross-sectoral law enforcement mechanism.

11.3.4 Consolidating Fundamental Work with Enhanced Monitoring and Surveillance

We will solidify fundamental work of water resources protection through investigation and review of basic information in water functional zoning, sewage outlets, and drinking water source areas in the basin, to improve basin information archive. With advanced technologies such as drone, remote sensing, IOT, video tracking and identification, monitoring network systems will be greatly enhanced. With mobile application tools, we will build a comprehensive monitoring and early warning system, which combines automatic and laboratory monitoring, and autonomous monitoring and information sharing. The above-mentioned will also promote the establishment of the emergency decision support system for water pollution accidents. All these measures will improve the comprehensive monitoring capacity on trans-provincial boundaries, water functional zoning, sewage outlets, drinking water source areas, ecological flow, and emergency monitoring.

We will advocate the improvement of water resources protection monitoring system, by implementing measure detection of sewage outlets, and increasing coverage of automatic online monitoring. Based on the national and local water resources management information systems, and combined with the long-term and dynamic management platform of rivers and lakes protection, we will establish the problem reporting system and problem listing management system, for water resources carrying capacity assessment, ecological flow ensure, drinking water source area safety evaluation, and red-line control of water-related ecological protection.

We will integrate existing water-related information in the basin under the MWR, and increase access to water-related data from other ministries of natural resources, ecological environment, agriculture, transportation, energy, and meteorology, so as to build up a big data center for water safety and water ecological rehabilitation in the Yangtze River Basin, based on which the open, sharing and mutually beneficial innovative mechanism and cooperation platform will be established.

11.3.5 Enhancing Scientific Research to Strengthen Capacity Building

More emphasis will be attached on research of key technologies and major issues, including carrying capacity of water resources, water environment and water ecology, control of pollutant into the river, drinking water sources security, and new technology of water environmental monitoring. We will promote research and application of scientific technology and standards for ensuring ecological flow of major rivers and lakes, regulation and control on pollutant carrying capacity of water bodies, and pollution risks control and emergency settlement. Platform development is also a key part, based on which we will encourage scientific and technological innovation in laboratories of research institutes under river basin organizations, field stations, remote sensing stations, and demonstration zoning, and will strengthen platforms to promote scientific research in supporting water resources protection, so as to enhance the supportive role played by science and technology development.

11.3.6 Addressing Watershed-Specific Rehabilitation Priorities

For major cities in the mainstream of the Yangtze River, we shall control the amount of pollutants discharged from major cities and improve the layout of water intakes and sewage outlets. Regarding major cities in the basin, such as Shanghai, Nanjing, Wuhan, Chongqing and Panzhihua, we should intensify efforts in protecting water resources, strictly implement plans to control the total amount of major pollutants entering the river, and actively promote the construction of water-conserving cities and water-ecological civilization cities.

For major tributaries and key areas, we will further improve comprehensive rehabilitation. Progress will be made in increasing urban sewage treatment rates along the Han River, Xiang River, Jialingjiang River, Tuojiang River, Min River and other major tributaries to reduce pollutant discharge. We will strengthen control of point source pollution in urban areas and non point source pollution in rural areas. At the same time, we will reinforce control of pollution caused by cadmium, mercury and other heavy metals in the Xiangjiang River, and prevent and control algal blooms in the middle and lower reaches of the Han River. We will strengthen ecological and environmental protection in Dongting Lake, Poyang Lake and other lakes, speed up the control of non-point sources pollution in Chaohu Lake and Dianchi Lake, carry out ecological rehabilitation of rivers flowing into the lakes to reduce the load of pollution, so as to ensure the ecological security of lakes. Other key areas, such as the source of the Yangtze River, the Three Gorges Reservoir area, the Danjiangkou Reservoir area, and the Yangtze Estuary, will also be a focus for water resources protection and restoration.

Chapter 12
Major Scientific and Technological Research Subjects

The Yangtze River Basin has large and varied tributaries, and the water resources very unevenly distributed, both in space and time. Floods and droughts occurred frequently in history. With rapid socio-economic development, ever-growing urbanization and intensified impact of global climate change, the hydrological situation of the river basin has undergone dramatic changes. Meanwhile, natural evolution and human activities often interact on each other. These factors are posing new challenges to the management of water-related behaviors of human being and the scheduling of hydro projects. According to the requirements of ensuring the water security of river basins, it is necessary to carry out researches on key scientific and technological issues on river basin rehabilitation and protection based on the previous studies and by applying modern technologies, so as to effectively improve the technical and managing capacity of river basin and provide systematic and effective technical support.

In view of the prominent problems existing in river basin rehabilitation and protection, six key subjects on river governance are proposed from the following directions: (1) flood control, drought relief and disaster mitigation; (2) joint operation of key projects; (3) conservation and optimal allocation of water resources; (4) protection of water resources and ecological environment; (5) regulation and protection of river and lake; (6) comprehensive river basin management.

© Changjiang Press (Wuhan) Co., Ltd. 2022
Changjiang Water Resources Commission of Ministry of Water Resources of the People's Republic of China, *Report on Yangtze River Rehabilitation and Protection 2019*,
https://doi.org/10.1007/978-981-16-4927-1_12

12.1 Flood Control, Drought Relief and Disaster Mitigation

12.1.1 Research on Multi-scale Hydrometeorological Forecasting Technology Under the Changing Environment

Accurate and timely hydrological and meteorological forecasting information is the "eyes" and "ears" for flood control, drought relief and disaster mitigation. In recent years, with the rapid development in hydrometeorological monitoring, hydrometeorological coupling and other aspects, hydrometeorological forecasting technology has achieved considerable progress. However, global climate change and operation of huge group reservoirs in the river basin have brought new challenges to hydrometeorological forecasting and put forward higher requirements. Therefore, it is urgent to carry out research on multi-scale hydrometeorological forecasting technology under changing environment, to extend the forecasting period, increase the forecasting accuracy, and ultimately give full play to the flood regulating and drought mitigating ability of hydro projects.

The research shall be carried out in the aspects of quantitative precipitation forecasting method for isolated heavy rainfall process, multi-scale rainfall numerical forecasting method and application, interval flood forecasting technology based on flood control compensation operation of reservoirs, evolution rule of reservoir flood wave in heavily modified rivers, and probabilistic flood forecasting method based on hydrological and meteorological coupling.

12.1.2 Research on Flood Management of the Yangtze River

Presently, the Yangtze River flood control is in the transitional period from "water adapting to human" to River flood control is in the Embankments along the mainstream and the major cities in the middle and lower reaches of the Yangtze River have reached the flood control standard. A large number of key reservoirs such as the Three Gorges have been built on the mainstream and the tributaries. The flood control capacity of the whole river basin has been greatly enhanced. Measures of removing river-encroaching polders for flood discharge, returning reclaimed lands to lake, and resettling local residents to restore room for floods have been implemented in the middle and lower reaches of the Yangtze River. Positive exploration and practice have been carried out in flood management, including risk management of flood storage and detention areas, flood utilization through reservoirs scheduling, flood hazard mapping, etc. In the New Era, flood control and disaster mitigation of the Yangtze River will undertake in accordance with the concept of human and water in harmony, and meet the overall requirements of "Two Insists and Three Transformations" and change our mindset of flood control, which requires in-depth research on flood management, so as to provide technological support for the transformation from "flood control" to "flood management".

The research shall focus on aspects of current situation and problem analysis of flood management, flood risk identification, flood management laws and regulations, flood management system and mechanism, flood management guidance of the Yangtze River.

12.1.3 Appraisal of Flood Control Layout Adjustment in the Middle and Lower Reaches of Yangtze River Basin

Flood control water level is the controlling safety index of the middle and lower reaches of the Yangtze River, and is the basis for flood control layout. With further development of the Yangtze River Economic Belt, flood control protection areas have higher requirements for flood control security. The river–lake relation and the watershed hydro project status have changed significantly. The conflict between human being and water has become increasingly prominent in some regions especially in the flood storage and detention areas. It is hard to make decisions whether to use flood storage and detention areas in case of emergency. The flood detention area would suffer huge losses once been used, which has become the main weakness of flood control construction in the middle and lower reaches of the Yangtze River. It is urgently needed to conduct the appraisal of flood control layout adjustment in the middle and lower reaches of the Yangtze River Basin, so as to coordinate the interests and relationships between upstream and downstream, left and right banks, main streams and branches and among different regions, to further improve flood control system in the middle and lower reaches of the Yangtze River, to guide the construction of flood control projects such as embankments of the middle and lower reaches of the Yangtze River, flood storage and detention areas, and to provide technical support for a new round of revision for Flood Control Planning of Yangtze River Basin and Master Plan of Yangtze River Basin.

The research shall focus on aspects of changes of the relationship between rivers and lakes in the middle and lower reaches of the Yangtze River under the new flow-sediment conditions and regulation effect of key projects, flood storage and diversion effect of floodplains and polders in the middle and lower reaches of the Yangtze River, the flood control water level in the middle and lower reaches of the Yangtze River, construction and management scheme of flood storage and detention areas in the middle reaches of the Yangtze River, and optimized layout scheme of flood control in the middle and lower reaches of the Yangtze River, etc.

12.1.4 Research on Extreme Floods and Droughts Under Climate Change and Countermeasures

The Yangtze River Basin is a strategic water source in water resources allocation of China and an important support for sustainable socio-economic development of the Basin and even the whole country. Water resources are relatively abundant in the Yangtze River Basin, but the spatial and temporal distribution of precipitation is uneven, which is vulnerable with frequent occurrence of floods and droughts. With the deepening impact of climate changes, water cycle elements (such as precipitation, evaporation, runoff, soil moisture, etc.) have undergone dramatic changes, leading to redistribution of water resources in time and space, resulting in widespread and frequent occurrence of extreme hydrological events such as droughts and floods in the basin. It is necessary to study the evolution characteristics and developing trend of extreme flood and drought events considering future climate changes, and put forward comprehensive countermeasures, which can help ensure water supply and improve integrated management of the Basin.

The research shall focus on aspects of spatial–temporal evolution characteristics and driving mechanism of extreme floods and droughts in the Yangtze River Basin, floods and droughts evolution prediction under climate changes in the Yangtze River Basin, mountain torrent disaster prevention and control technology and demonstration in the Yangtze River Basin, the study of the glacier lake generation, disappearance and disaster prevention in typical area of Tibet under climate changes, extraordinary flood defense strategy in the Yangtze River Basin, control and operation technology of emergency water resources under the condition of extreme droughts in the Yangtze River Basin, etc.

12.1.5 Research on Risk Zoning of Dammed Lake Disaster on the Upper Reaches of the Yangtze River and the Relevant Countermeasures

Quite a few parts of the upper reaches of the Yangtze River are located in the first and second grade terraced and transitional parts of China's terrain, where various geological disasters occur frequently, dammed lake events take place rather often. Particularly, the 2008 earthquake dammed lake and Baige Dammed Lake event on the Yangtze River upper reaches (in Jinshajiang river) from October 2018 to November 2018 had seriously threatened to the lives and properties near and along the downstream area of the reservoir, which made great challenge to the operation of the reservoir, aroused great concerns from governments at all levels and the whole society. However, due to uncertainty of the geological conditions, incomplete hydro-met data, inconvenient land and water transportation, short response time and other factors, while considering limitation from risk management, response mechanism, disposal technology and equipment system, the disaster control ability of dammed lake is relatively weak, which may lead to disastrous consequences. In order to mitigate the impact of the disasters, it is necessary and urgent to carry out the research on dammed lake disaster risk zoning and control strategy on the upper reaches of the Yangtze River.

The research shall focus on aspects of investigation of geological disaster development characteristics, formation mechanism of dammed lake, risk zoning of dammed lake disaster, dam break flood prediction technology of dammed lake, disaster prevention and mitigation strategy of dammed lake, etc.

12.2 Conservation, Utilization and Optimized Allocation of Water Resources

12.2.1 Research on Impacts of Inter-basin Water Transfer Projects on the Yangtze River Basin and Risk Prevention and Control Technology

The Middle Route and the East Route of the South-to-North Water Transfer Project have been implemented, and the projects of Water Diversion to Central Yunnan, Yangtze-to-Huaihe Water Diversion are under construction. The West Route of the South-to-North Water Transfer Project is still in the preliminary stage. In the process of planning, implementation of the inter-basin water transfer projects, a large number of scientific researches, planning, and feasibility studies have been conducted, which mainly focus on a single water transfer project. Few Studies have been conducted to ascertain the comprehensive effects of multiple inter-basin water division projects on the whole Yangtze River Basin. It is lack of long-term stable compensation mechanism for the water source region. Systematic risk assessment and prevention and control measures have not been formed for the inter-basin water transfer projects. Research on the superposition impacts of inter-basin water transfer projects on the Yangtze River Basin and risk prevention and control technology can provide technical support for the planning, design, construction and operation of the subsequent inter-basin water diversion projects, and establishment of the compensation mechanism of the inter-basin water diversion projects for the water source regions, and can also provide basis for risk prevention of the inter-basin water diversion projects.

The research shall focus on aspects of status quo investigation of the inter-basin water diversion projects of the Yangtze River, impact of the inter-basin water diversion project on the Yangtze River Basin, compensation mechanism system of water source regions of the inter-basin water diversion projects, risk assessment system, prevention and control measures of the inter-basin water diversion projects.

12.2.2 Research on Coordinated Management and Control of Water Resources for Ecological Protection in the Hanjiang River Basin

In October 2018, the State Council approved the official document of the Development Plan of the Hanjiang River Ecological Economic Belt referred to (2018) No. 127. Under the premise of ecological conservation, how to coordinate the flood control, water supply, power generation, navigation and other tasks in Hanjiang River basin, so as to achieve coordinated development of quantity, quality and efficiency of water resources, promote construction of the Hanjiang River Ecological Economic Belt, and realize efficient utilization and efficient management of water resources in the Hanjiang River basin, is one of the key issues for the Hanjiang River basin.

Through analyzing ecological conversation requirements of the Hanjiang River Basin in response to new situation, and on the basis of the requirements of existing planning and related studies, the research shall focus on aspects of ecological conservation requirements and existing problems in the Hanjiang River Basin, the water demand of the Hanjiang River Basin in the target year of planning in 2035 under the ecology priority conditions, optimal allocation scheme of water resources of the Hanjiang River Basin in the target year of planning in 2035, coordinated control scheme and measures of water resources for ecological conversation in the Hanjiang River basin, dynamic response and joint operation of Danjiangkou Reservoir for water quality, water quantity and pollutants, etc.

12.2.3 Research on Water-Saving Technologies in Large-Scale Irrigation Districts in the Yangtze River Basin

The irrigation water use accounts for more than 50% of the total quantity of water uses in the Yangtze River Basin. Agriculture is still the largest water consumer and serves as the industry with the most potential of water saving. By 2015, the large-scale irrigation districts which had completed engineering corollary and water-saving reconstruction only accounted for about 20% in Yangtze River Basin. It is essential to vigorously develop water-saving irrigation, popularize advanced water-saving irrigation scheduling, improve irrigation assurance rate and water use efficiency, strengthen the management of water demand, and accurately predict water-saving potential of irrigation districts by obtaining the field water use data through the construction of informatization system, so as to protect and improve the ecological environment of irrigation districts, meet the requirements of efficient water use and serve the strategy of rural revitalization, guarantee food security and realize the agricultural modernization.

The research shall focus on aspects of establishing precision irrigation model, realizing efficient water-saving irrigation model, study of new irrigation pattern with integrated service function, study of water dispatching system based on automatic control technology, demonstration of water-saving technology application in typical irrigation districts, etc.

12.2.4 Research on Key Technologies for Controlling and Managing Total Quantity of Water Uses in the Yangtze River Basin

The total quantity of water uses is one of the Three Red Lines of the strictest possible systems for water resource management, and it is the key to ensure that the scale of water resources development and utilization fits the carrying capacity of water resources and water environment. At present, all provinces (including autonomous regions and municipalities directly under the central government) in the Yangtze River Basin have basically completed the decomposition of the total quantity of water uses indexes in the administrative areas at the provincial, municipal and county levels, and have formulated water allocation plans. These plans are for the important trans-provincial rivers such as Hanjiang, Jialinjiang, Minjiang, Tuojiang, Chishuihe, Wujiang, Niulanjiang, Jinshajiang, Yuanjiang, etc. However, the total quantity of water uses has not been fully decomposed into rivers, reaches, water sources and water-used industries. The constraint driving mechanism of the total quantity of water uses on the socio-economic development planning has not been established. It is still lacking the technical support for dynamic control and the appraisal system of the total quantity of water uses as well as a forceful regulatory instrument for controlling the total water consumption. It is urgent to carry out the research on key technologies for controlling and managing total quantity of water uses in the Yangtze River Basin to realize red-line control of water resources development and utilization and dual control of total quantity and intensity of water resources consumed.

The research shall focus on aspects of detailed scheme of the total quantity of water uses in the river basin based on optimal allocation of water resources, key technologies of dynamic control of water intake and use in the basin based on the total consumption control index, and the key technology of dynamic monitoring and regulation of carrying capacity of water resources in the river basin based on the background of well-coordinated environmental conversation of the Yangtze river, etc.

12.3 Protection of Water Resources and Eco-environment

12.3.1 Research on the Eco-environment Evolution Rule and Protection Strategy of the Rivers and Lakes in the Source Region of the Yangtze River

The source region of the Yangtze River is located in the hinterland of the Tibet Plateau, which is an important national ecological security shelter. In the recent 50 years, due to the dual impact of global climate change and human activities, aquatic environment in the source region of the Yangtze River has been deteriorating, directly threatening the stability of the ecosystem in the source region of the Yangtze River. At present, scholars from home and abroad have carried out researches on ecological environment protection in the source region of the Yangtze River, including the relationship between plateau climate and glacier change, response of plateau lake wetland evolution to climate change, impact of climate change on the ecological environment in the source region and other related aspects. But due to the poor natural conditions, traffic inconvenience, cold and hypoxia circumstance in the source region of the Yangtze River, so far it lacks basic data on the eco-environment evolution rule and driving mechanism of the rivers and lakes in the source region of the Yangtze River. It is necessary to conduct a comprehensive, systematic and in-depth research to provide basic data and decision-making basis for eco-environment protection and ecological risk prevention and adaptive management of rivers and lakes in the source region of the Yangtze River under climate change.

As for the source region of the Yangtze River, the research shall focus on aspects of investigation and evaluation of aquatic ecological environment current status of rivers and lakes, water cycle regularity and mechanism, ecological hydrology change rule and aquatic biological protection strategy under climate change, basic rule and protection strategy of water–sediment transportation and riverbed evolution, wetland ecological evolution and protection strategy, response and ecological regulation of the alpine grassland ecosystem to human disturbance and climate change, evolution mechanism of the main plateau lakes, etc.

12.3.2 Research on Management and Technology of Water Quality Safeguards for Important Drinking Water Sources in the Yangtze River Basin

A total of 221 water sources in the Yangtze River Basin were listed as important drinking water sources in 2016, accounting for 36% of the total. But heavy chemical enterprises are densely covered, 30% of which are located within 5 km around the drinking water source with environmental risk. Main drinking water sources are located alternately with production, storage and transportation concentration areas of various kinds of dangerous and heavy pollution sources. The threat from crossing traffic facilities and moving pollution risk sources including dangerous chemical transport vessels and vehicles still exists in the water source protection zones. The responsibilities of local governments in the construction and protection of drinking water sources are not clear enough, and some water sources have not been completely isolated. Online monitoring facilities have not been built in some water sources protection zones, and regular monitoring of the overall 109 indicators have not been conducted in some water sources protection zones. The installation rate and standardization level of monitoring facilities still need improving. Therefore, it is urgent to carry out research on management and technology of water quality safeguards for important drinking water sources in the Yangtze River Basin, so as to provide important guarantee of people need improving. Therefore, it is urgent to carry out research on management and technology of water quality safeguards for important drinking water sources in the Yangtze River Basin, so as to provide important safeguards of people's livelihood for the construction of the Yangtze River Economic Belt.

The research shall focus on aspects of assessment of the current security status and the security management mechanism of the important drinking water sources in the Yangtze River Basin, investigation and research on the water quality safety risk of the typical channel water sources, key technology of the water quality safeguards for important drinking water sources, etc.

12.3.3 Research on Key Technology of Water Function Zone Management in the Yangtze River Basin

Water function zone management is the core of water resources protection system. Water quality compliance rate in water function zone is one of the Three Red Lines of the strictest possible systems for water resource management. According to the Water Function Zoning for the Major rivers and Lakes of China approved by the State Council in 2011, a total of 1521 water function zones and their water quality management targets have been delimited for the Yangtze River Basin and the Southwest rivers. But in the practice of the management, there are still some problems as following: the restraint is insufficient for the total pollutant discharge quantity control in water function zones; procedures and technical requirements for water function zoning adjustment are incomplete and are difficult to implement; further review and adjustment are urgently needed on the pollutant assimilation capacity and the total quantity of the pollutant discharge in water function zones; further investigation and review are needed for the monitoring sections of water function zones, water quantity control and ecological water requirements are not included in the targets of water function zone management. Therefore, it is urgent to carry out research on the key technologies of water function zone management.

The research shall focus on aspects of water function zoning technology and sustainable utilization mode of water function, study on impact factors of the water pollutant assimilation capacity of water function zones and the Red Line management, dynamic total pollutant discharge control technology based on the carrying capacity of the aquatic environment in the river basin, evaluation and assessment system for water function zones based on the river health, equipment integrated technology of real-time color intelligent identification and early warning system for sewage outlets in the Yangtze River Basin, etc.

12.3.4 Deepening Research on Identification of Ecological Demands and Determination of Ecological Flow of Major Rivers and Lakes in the Yangtze River Basin

The amount of water resources, hydrological processes, development and utilization degree of water resources and water quality in different regions of the Yangtze River Basin vary greatly. The water ecological system structure and function, the objective of ecological protection of different regions are also different. Hence, there are obvious differences on ecological requirements and ecological flow target among different rivers, the applicability of the existing methods needs studying. On the basis of previous research results, it is necessary to carry out deepening research on identification of ecological demands and determination of ecological flow of major rivers and lakes in the Yangtze River Basin, so as to provide scientific basis for the formulation of ecological flow, and provide support for supervision, management and safeguard of ecological flow.

The research shall highlight on aspects of the accounting method and accounting technical system of ecological flow of rivers and lakes, the formulation of the ecological flow target, the Red Line control of ecological flow and its assessment mechanism, the ecological water demand analysis in ecological sensitive areas, etc.

12.3.5 Research on the System of Securing Ecological Water Quantity for Major Rivers and Lakes in the Yangtze River Basin

The ecological water flow assurance of rivers and lakes is the basic requirement for strengthening the management and control of water resources development and utilization and promoting rivers and lakes ecological protection and rehabilitation, and matters the broad picture of ecological civilization and water conservancy reform and development. So, it is urgent to carry out research on the system of securing ecological water quantity of major rivers and lakes in the Yangtze River Basin.

The research shall focus on aspects of the monitoring and early warning mechanism of ecological flow as well as the ecological water demand zoning and section control system in major rivers and lakes in the Yangtze River Basin, theory and method of optimal allocation of water resources in the river basin based on ecological flow, regulation and technological standard system of ecological water, etc.

12.3.6 Research on Water Ecological Impact of Cascade Group Reservoirs in the Yangtze River Basin and Mitigation Measures

In the Yangtze River Basin, 329 large reservoirs have been built, with a total storage capacity of more than 180 billion m^3 and a flood control storage capacity of about 77 billion m^3, of which 40 reservoirs have been included in the joint operation of reservoirs of the Yangtze River flood control. Cascade reservoirs play an important role in flood control, power generation, navigation and water resources allocation, but they also have impact on aquatic habitat and hydrological regime. In order to fully implement the well-coordinated Yangtze River Conservation and ensure the comprehensive benefits of cascade reservoirs, it is necessary and urgent to carry out in-depth research on water ecological impact of cascade group reservoirs and mitigation measures.

Taking the cascade reservoirs on the middle reaches of Jinsha River, the upstream of Yangtze River, the middle and lower reaches of Hanjiang River, and the lower reaches of Wujiang River as the key points, the research shall focus on aspects of monitoring and evaluation system of ecological effect of cascade reservoirs, aquatic habitat evolution and habitat suitability assessment of cascade group reservoirs on the Yangtze River upstream, running water habitat shrinking ecology effect and response mechanism of important aquatic organisms, water temperature monitoring and ecological effect assessment of the key reservoirs on the main stream of the Yangtze River, the supersaturation ecological effect of discharge gas from cascade hydropower stations in the Yangtze River Basin and its mitigation countermeasures, the key technologies of multi-objective fish migration channel recovery in the Yangtze River Basin, important aquatic organisms artificial propagation and releasing management and effect evaluation in the Yangtze River Basin, etc.

12.3.7 Research on Key Technologies of Soil and Water Conservation in the Yangtze River Basin

It is necessary to carry out researches on key technologies of soil and water conservation as the following outstanding problems existed. Long assessment period of the soil and water losses status in the Yangtze River Basin, unrefined grasp of soil and water losses information, imperfect system of responsibility target management on soil and water conservation technical support, serious water and soil erosion in the slope croplands, and insufficient research on typical fragile areas ecosystem monitoring and ecological restoration technology.

The research shall focus on aspects of rapid monitoring technology system of soil and water losses in the river basin, assessment of ecological environment status quo and responsibility target assessment index of soil and water conservation, nature rehabilitation management and technology system for slope croplands, ecosystem degradation pattern and trend monitoring, etc.

12.4 Regulation and Protection of Rivers and Lakes

12.4.1 Research on River Channels Evolution and the River–Lake Relation Changes on the Middle and Lower Reaches of the Yangtze River Under the New Flow-Sediment Conditions

The river channels evolution and the river–lake relation changes on the middle and lower reaches of the Yangtze River under the new flow-sediment conditions are important issues to be studied and also are the basis for the study of flood control, water utilization, water environment and water ecology protection and so on under the new flow-sediment conditions on the middle and lower reaches of the Yangtze River. In recent years, with the flow-sediment regime featuring significant reduction of sediment discharge on the middle-lower Yangtze River, the whole mainstream river channel on the middle and lower reaches are eroded. At the same time, river regime in some reaches adjust to the changes dramatically; exploitation and protection of the shoreline and promotion of the gold waterway capacity are restricted; flow-sediment conditions, degradation and aggradation in rivers and lakes and the resulting river–lake relationship changes are long-term and complicated process. On the basis of previous research results, it is necessary to further strengthen the comparative analysis of the latest measured data, and continue carrying out in-depth research on erosion and deposition evolution of river channels on the middle and lower reaches of the Yangtze River and changes of the river–lake relation under the new flow- sediment conditions, so as to provide scientific basis for regulation and protection of rivers and lakes on the middle and lower reaches of the Yangtze River under new circumstances.

The research shall focus on aspects of the change and trend of water and sediment, observation and research on the change of river degradation and aggradation on the middle and lower reaches of the Yangtze River, the change of the stage-discharge relation in the main stations on the middle and lower reaches of the Yangtze River, the change of the relationship between the Yangtze River and the Dongting Lake, and the change of the relationship between the Yangtze River and the Poyang Lake, etc.

12.4.2 Research on River Regime Monitoring, Bank Collapse Warning and Regulation Technology on the Middle and Lower Reaches of the Yangtze River

The middle and lower reaches of the Yangtze River is alluvial plain river, and the geological structure of the bank is mostly in dual structure. The collapse hazard often occurs in the concave bank facing into the major current, and the bank collapse has the characteristics of sudden, multiple and random. Since the founding of the People's Republic of China, continuous efforts have been made to protect embankments, control river regime and provide emergency rescue, but the traditional riprap revetment adopted in bank collapse regulation will damage the natural ecology of mountain. So it is urgent to optimize and develop a new technology for bank collapse regulation, evaluate on bank slope stability risk and timely forecast the possibility of bank collapse, in order to improve the emergency response capability and implement regulation timely, which will provide necessary technical support for the flood control on the middle and lower reaches of the Yangtze River.

The research shall focus on aspects of the mechanism of bank collapse, river regime monitoring technology, early-warning technology of bank collapse, regulation technology of bank collapse, emergency rescue mechanism of bank collapse on the middle and lower reaches of the Yangtze River, etc.

12.4.3 Research on Evolution Response Mechanism and Regulation Technology of Key Reaches in the Middle and Lower Reaches of the Yangtze River

After the operation of the Three Gorges Reservoir and the upstream reservoirs, less sediment laden flow has being discharged for a long period of time, which caused long distance degradation of the middle and lower reaches of the Yangtze River, bank collapse in some part of the reaches, flow adjustment in braided rivers, constantly degradation and aggradation revolution on reaches with river islet densely covered and dramatic change in river regime. At the same time, in the past, impermeable and hard revetment were often used in river regulation projects, which cut off the organic connection between terrestrial ecosystem and water ecosystem, leading to the capacity weakening of self-purification of water and water ecosystem resilience. As a result, it is necessary to carry out the ecological rehabilitation and it is urgent to form a set of methods to guide the engineering application of river ecological rehabilitation.

The research shall focus on aspects of evolution of the key reaches on the middle and lower reaches of the Yangtze River after the operation of the Three Gorges Project, impact of the key reaches evolution on water resources and ecological environment on the middle and lower reaches of the Yangtze River, damaged river slope ecological habitat restoration technology, embankment ecological reconstruction scheme for the ecological sensitive region on the middle and lower reaches of the Yangtze River, river regulation strategy for the key reaches on the middle and lower reaches of the Yangtze River, etc.

12.4.4 Research on Comprehensive Regulation and Protection of Dongting Lake and Poyang Lake Under Changing Environment

In recent years, due to the impacts of natural geographical conditions, human activities (especially the joint operation of large-scale water conservancy projects) and climate change, extreme hydrological events happen frequently, and the river–lake relations change significantly. In view of water resources, flood control, water ecological environment, navigation and other related issues of the Dongting Lake and Poyang Lake, in-depth study of comprehensive regulation and protection scheme of the Two Lakes under changing environment is of great significance to gradually improve the diversion capacity of the Three River Mouths on Jingjiang reaches of Yangtze River, realize win–win for both rivers and lakes, effectively protect the clear water of lakes, and promote the development of the Yangtze River Economic Belt.

The research shall stress aspects of flood and drought disaster causes and evolution rule under changing environment in Dongting Lake and Poyang Lake, scheduling of sluices in lakes and rivers, interconnection of river–lake system and floodway regulation, aquatic ecology and environment change in Dongting Lake and Poyang Lake, comprehensive water pollution control and ecological environment protection system of Dongting Lake, aquatic ecological environment protection and restoration of Poyang Lake, wetland conservation and ecological hydrology threshold optimization for Dongting Lake and Poyang Lake, etc.

12.4.5 Research on Protection and Management of Floodplains in the Middle and Lower Reaches of the Yangtze River

There are a large number of floodplains on the middle and lower reaches of the Yangtze River, which play important role in flood storage, river regime stability and ecological protection, and are important part of the flood control system on the middle and lower reaches of the Yangtze River. Research on protection and management of floodplains on the middle and lower reaches of the Yangtze River is an inevitable choice to ensure the play of flood storage, an objective requirement to reduce flood storage loss, an urgent need to regulate the development and utilization of floodplains, and an objective need and important measure to develop the Yangtze River Economic Belt in the protection.

The research shall underline aspects of classification of floodplains and polders, population resettlement plan of floodplains and polders, ecological rehabilitation and protection plan of floodplains and polders, classified management idea for floodplains and polders, etc.

12.4.6 Research on Long-Term Evolution Process and Comprehensive Regulation of the Main Channel and Floodplain of the Yangtze River Estuary

The Yangtze River estuary is in the "leading" position of the construction of the Yangtze River Economic Belt, which occupies a pivotal position in the socio-economic development of the Yangtze River Basin and even the nation. In recent years, due to the comprehensive influence of hydrological and meteorological condition changes, runoff and storm surges, the operation of upstream water conservancy projects and human activities, the dynamic flow-sediment process in the Yangtze River estuary is complex and changeable, and the river regime in the Yangtze River estuary has undergone new changes. The river regime and stability of the main channel and floodplain of the estuary concerns the flood security, water safety, navigation safety, ecological environment protection of the cities along the Yangtze River. Therefore, conducting systematical researches on long-term evolution process, the driving mechanism and the effect of the ecological environment of the channel and floodplain of the Yangtze River estuary and proposing the comprehensive regulation idea in consideration of the ecological environment needs can provide scientific basis for protection and management of the Yangtze River estuary and the revision of the comprehensive development planning under the new circumstances.

As for the Yangtze River estuary, the research shall focus on aspects of the temporal and spatial changes of the water, sediment and salinity, the stability of the main channel and floodplain under the new flow-sediment conditions, the ecological environment effect of the long-term evolution of the main channel and floodplain, the ideas of the comprehensive protection and regulation, etc.

12.5 Joint Operation of Key Projects

12.5.1 Research on Ecological Operation of Key Group Reservoirs in the Yangtze River Basin

It has accumulated some experience and made some achievements on ecological regulation of reservoirs in the Yangtze River Basin, but the research foundation remains weak. Except for the natural propagation of the Four Famous Domestic Chinese Fishes, the scheduling index and the feasibility study for other important species, important ecological function areas, wetland protection and restoration of aquatic environment are in the initial stage. The research on the key technology of and the impact of ecological operation of group reservoirs on the aquatic ecological environment in the river basin level also needs to be further strengthened. In order to fully implement the environmental conservation strategy of the Yangtze River, guarantee the basic ecological water use of rivers and lakes, and mitigate the impacts of cascade reservoirs development and operation on important aquatic organisms and ecological environment, it is necessary and urgent to carry out in-depth research on the ecological operation of group reservoirs in the Yangtze River Basin.

The research shall focus on aspects of ecological operation needs, multi-objective scheduling coordination game of key group reservoirs of the Yangtze River for ecology, power generation, water supply, navigation and so on, operation strategy of group reservoirs with ecological requirements on the downstream of the Jinshajiang River and the Three Gorges Reservoir, ecological operation of reservoirs and harnessing of algal bloom on the middle and lower reaches of the Hanjiang river, joint ecological operation of key group reservoirs of the Yangtze River Basin, etc.

12.5.2 Further Research on Joint Operation of Key Group Reservoirs on the Upper and Middle Reaches of the Yangtze River

With the gradual regulation and development of the Yangtze River Basin and the construction of a group of water conservancy projects such as Baihetan, Wudongde, and the boundary conditions of the joint operation of group reservoirs have changed, the cumulative impacts of construction and operation of projects on the river basin are gradually appearing. It is urgent to further deepen the research on the joint operation of key group reservoirs on the upper and middle reaches of the Yangtze River, to coordinate various needs from each side, relieve the competition between different objectives, so as to adapt to the new management requirements of river basin under the new circumstances, and give full play to the comprehensive utilization benefits of group reservoirs on the upstream of the Yangtze River.

The research shall focus on aspects of joint flood control operation of the four reservoirs on the downstream of Jinshajiang River and the Three Gorges Reservoir, coordinated flood control operation of key group reservoirs on the upper and middle reaches of the Yangtze River, joint benefit promotion operation of key group reservoirs on the upper and middle reaches of the Yangtze River, joint sediment regulation of key group reservoirs and integrated regulation mode of group reservoirs on the upstream of the Yangtze River, etc.

12.5.3 Research on Joint Operation of Group Reservoirs and Other Key Projects in the Yangtze River Basin

The joint operation of group reservoirs for the requirements of flood control, ecology, electricity, water supply scheduling and so on, has been realized on the upper and middle reaches of the Yangtze River but has not met the requirements of flood control, waterlogged drainage and water supply on the middle and lower reaches. In order to give full play to a variety of water resources and hydropower projects in the Yangtze River Basin in floods and droughts mitigation, water resources allocation, water ecological environment protection, restoration and integrated river basin management, etc., on the basis of joint operation research on the key group reservoirs on the upper and middle reaches in the Yangtze River Basin, it is necessary to conduct research and practice on the integrated joint operation of the river basin key reservoirs and the other key projects to meet with the requirements of flood control, waterlogged drainage, water supply in the middle and lower reaches of the Yangtze River Basin.

The research shall focus on aspects of joint operation of key projects to guarantee the water supply security in the Yangtze River Basin, inter-basin water resources joint operation of key projects, flood control joint operation of group reservoirs and other key projects, etc.

12.5.4 Research on Management Mechanism and Risk Control of the Joint Operation of Group Reservoirs in the Yangtze River Basin

The group reservoirs of the Yangtze River subordinate to different administrative bodies and are managed by different units. There are problems such as decentralization of authorities and limitation of objectives, which may easily lead to problems such as difficulty in realizing benefits of some reservoirs, difficulty in effectively slowing down or even aggravating the adverse effects on water resources and water ecological environment in the basin. In order to ensure scientific and reasonable regulation and operation of group reservoirs in the basin, maximize their comprehensive benefits and minimize their adverse effects, it is necessary to gradually establish a unified regulation and management system of group reservoirs in the basin by improving laws and regulations, perfecting management system, establishing management mechanism and strengthening technical support.

The research shall focus on aspects of joint operation management system and mechanism of group reservoirs of the Yangtze River Basin, emergency plan of joint operation of group reservoirs to response to the major water safety incidents in the Yangtze River Basin, risk assessment and regulation technology of joint flood control regulation of group reservoirs on the upper and middle reaches of the Yangtze River, risk and benefit compensation mechanism of joint operation of group reservoirs on the upstream of the Yangtze River, risk control of joint operation of group reservoirs for the excessive flood, safety risk assessment and strategy research on the joint operation of large cascade group reservoirs, etc.

12.6 Integrated River Basin Management

12.6.1 Research on Shoreline Management of the Yangtze River Under the New Circumstances

Yangtze River shorelines are important part of the ecosystem. Since 2016, the Ministry of Water Resources and other ministries as well as other local governments along the river have carried out a series of special regulation actions, a large number of shoreline utilization projects have been rectified, standardized and upgraded, and the unreasonable utilization of shorelines in the Yangtze River has been significantly improved. But with the restriction of the socio-economic development stage and the limitation of historical condition, there are still some problems in the current protection and utilization of the Yangtze River shorelines such as unreasonable shoreline exploitation layout, extensive use, unauthorized tampering, lack of effective market, economic regulation and control management methods on shoreline resources development and utilization, which have restrained the effective protection and scientific utilization of shoreline resources and law-based management. With the advancing of the Yangtze River Economic Belt strategy, how to manage and protect the Yangtze River shoreline resources is an important task not only for present but also for a period ahead. Therefore, it is urgent to study the shoreline management of the Yangtze River.

The research shall stress aspects of protection and management objectives of the Yangtze River shoreline under the new circumstances, control strategy of the Yangtze River shorelines, the compensable use of the Yangtze River shorelines, information management system of the Yangtze River shoreline, management guidance for the Yangtze River shoreline, etc.

12.6.2 Research on River Chief System Management in the New Era

The River Chief System is a major institutional innovation to implement the new concept of green development, strengthen law-based water management, and promote the construction of ecological civilization legal system. The implementation of the River Chief System in the Yangtze River Basin, should be based on river basin integrity and inter-provincial numerous lakes, strive to resolve the uncoordinated contradiction between management in zoning and integrated river basin management of water-related affairs; explore to formulate the coordination and supervision plan which fit for the River Chief System; take an overall consideration for schemes, objectives, tasks, and strategies of River Chief System of different provinces and regions; prepare a coordination and supervision plan of River Chief System with clear responsibility, orderly coordination, strict regulation and strong protection, which are very urgent and significant.

The research shall focus on aspects of joint coordination mode, technical index system, and unified supervision mechanism of River Chief System in the Yangtze River Basin, etc.

12.6.3 Research on Sand Excavation Management Mechanism and Strategy in the Yangtze River Under the New Circumstances

In recent years, with successive construction and operation of group reservoirs on the upper reaches of the Yangtze River and the implementation of soil and water conservation projects, the amount of sediment coming from the upper reaches of the Yangtze River has been further reduced, and the degradation of the mainstream on the middle and lower reaches have developed to the downstream significantly, the sand and gravel on the middle and lower reaches of the river channel show an unidirectional shrinking trend, and the sand and stone demand and supply conflicts are increasingly prominent. It is urgent to study the sand excavation management mechanism and measures of the Yangtze River channel under the new circumstances in view of new flow-sediment change conditions, so as to provide technical support for promoting the development of river management in the new era, ensuring the socio-economic harmony and stability of the Yangtze River Economic Belt, and guaranteeing the long-term orderly operation of sand excavation management of the Yangtze River.

The research shall focus on aspects of investigation and analysis of sand and gravel resources distribution in the Yangtze River channel, overall evaluation of sand excavation management in the Yangtze River channel, development and utilization technology of sand and gravel resources in reservoir areas and the downstream of the dams on the upper reaches of the Yangtze River, the long-term mechanism index system and strategy of sand and gravel excavation management in the Yangtze River channel, etc.

12.6.4 Research on Mechanism of Payment Based on Ecosystem Service in the Yangtze River Basin

There is a lack of effective coordination and communication mechanism among different regions in the Yangtze River Economic Belt. Actively building an inter-provincial mechanism of payment based on ecosystem service flow will help to further clarify and straighten out the ecological compensation relationship among the provinces located in the upper, middle and lower reaches within the regions.

The research shall focus on the following aspects of the formation mechanism and quantitative evaluation method of ecosystem service flow in the Yangtze River Basin: the range and framework system of payment for water ecological services based on ecosystem service flow, mode and standard method of payment for water ecological services based on ecosystem service flow, scheme of payment for water ecological services based on ecosystem service flow, etc.

12.6.5 Research on Overall Framework and Key Technologies of the Smart Yangtze River

The long-term practice of river management shows that it is difficult to deal with the increasingly complex water problems effectively by only structural measures without informatization management. Only by widely applying modern information technology, centering on the four major problems including water disasters, water resources, water environment and water ecology, integrating informatization into all aspects of water conservancy work, and integrating informatization into the whole process of water conservancy business, can the transformation and upgrading of traditional water conservancy industry be driven and the water conservancy modernization of the Yangtze River be realized. Therefore, it is urgent to carry out research on the overall framework and key technologies of the Smart Yangtze River System, so as to provide decision-making reference for the construction of the Smart Yangtze River System, etc.

The research shall focus on aspects of overall framework of Smart Yangtze River System, sharing mechanism of integrated information management in the Yangtze River Basin, construction of digital parallel space in the Yangtze River Basin, the smart hydrological monitoring system, etc.

Postscript-Words from the Translators

The Changjiang Water Resources Commission (CWRC) was established in February 1950, responsible to the Ministry of Water Resources (MWR) for integrated governance of the Yangtze River Basin and the Southwestern river basins. Under the leadership of the MWR, remarkable achievements have been accomplished and a wealth of expertise accumulated in rehabilitation and protection of the Yangtze River Basin in the past 70 years.

In recent years, the CWRC, taking advantage of its strengths and expertise in water resources and river basin management, has been proactively involved in international cooperation under the frameworks of the Belt and Road Initiative as well as bi-lateral and multi-lateral agreements in partnership with governments and entities of over 50 countries and regions all over the world. The international community recognizes the CWRC's contributions to building up the knowledge base of water resources management and eco-environment rehabilitation of river basins, and requests to share, in English version, solutions to the Yangtze River.

The Report that CWRC prepared in its original Chinese version provides a distilled summary of 70-year experience on the governance of the Yangtze River, as a result of painstaking efforts by several generations of engineers with the CWRC. We expect that the English edition of the Report could help those interested in the Yangtze River management better understand what has been done and what is going on in the Basin, especially with respect to the rehabilitation and protection of the third largest river of the world.

In the course of translation, we sincerely thank the Department of International Cooperation, Science and Technology of the MWR for their invaluable guidance. Our cordial acknowledgement is also extended to the Bureau of Programming and Planning and other departments of the CWRC for their facilitation and generous help.

As masterminded and organized by Mr. LUO Xiaoyong, the translation from Chinese to English of the Report was a joint effort by the staffs with the Bureau

© Changjiang Press (Wuhan) Co., Ltd. 2022

Changjiang Water Resources Commission of Ministry of Water Resources of the People's Republic of China, *Report on Yangtze River Rehabilitation and Protection 2019*,
https://doi.org/10.1007/978-981-16-4927-1

of International Cooperation, Science and Technology of the CWRC. The external experts, Mr. HUANG Jianhe and Mr. Jeff Kerr, separately reviewed the entire translated texts with corrections made after Mr. WANG Zhujun, Ms. Yang Lin and Ms. Liu Yuan edited all the translated parts. The Preface and Part One were translated by Ms. YANG Lin and proofread by Mr. YU Jicheng, Ms. LIU Yuan and Mr. WANG Zhujun. For the Part Two, Chap. 7 was translated by Mr. YU Jicheng and proofread by Ms. ZHOU Dongni; Chap. 8 translated by Ms. ZHOU Zhulin and proofread by Ms. ZHOU Dongni; Chap. 9 translated by Ms. ZHOU Dongni and Ms. LIU Yuan and proofread by Mr. WANG Zhujun; Chap. 10 translated by Mr. YU Jicheng and Mr. WU Quanfeng and proofread by Mr. WANG Zhujun and Mr. CHEN Li; Chap. 11 translated by Ms. LIU Lian and proofread by Ms. YANG Lin; Chap. 12 translated by Ms. ZHOU Dongni and proofread by Ms. YANG Lin.

In order to get the translated version out at an earlier date, the translators worked very hard under a tight schedule. However, the contents of the Report are rather extensive in coverage of almost all aspects of river basin management and are therefore multi-disciplinary, which posed a great challenge for the translators. We cannot rule out the presence of errors and inaccuracies in either interpretation or paraphrasing of the original Chinese version, for which your discretion and correction are welcomed.

Bureau of International Cooperation, Science and Technology.
Changjiang Water Resources Commission of Ministry
Resources of the People's Republic of China.

August, 2019.

The manufacturer's authorised representative in the EU is Springer Nature Customer Service Centre GmbH, Europaplatz 3, 69115 Heidelberg, Germany. If you have any concerns regarding our products, please contact ProductSafety@springernature.com

Printed and bound by CPI Group (UK) Ltd, Croydon, CR0 4YY
30/09/2026
02225287-0001